16 Cases of Mission Command

General Editor
Donald P. Wright, Ph. D.

Combat Studies Institute Press
US Army Combined Arms Center
Fort Leavenworth, Kansas

Foreword

For the US Army to succeed in the 21st Century, Soldiers of all ranks must understand and use Mission Command. Mission Command empowers leaders at all levels, allowing them to synchronize all warfighting functions and information systems to seize, retain, and exploit the initiative against a range of adversaries.

This collection of historical vignettes seeks to sharpen our understanding of Mission Command philosophy and practice by providing examples from the past in which Mission Command principles played a decisive role. Some vignettes show junior officers following their *commander's intent* and *exercising disciplined initiative* in very chaotic combat operations. Others recount how field grade officers *built cohesive teams* that relied on *mutual trust* to achieve key operational objectives.

Each historical account is complemented by an annotated explanation of how the six Mission Command principles shaped the action. For this reason, the collection is ideal for leader development in the Army school system as well as for unit and individual professional development.

Mission Command places great responsibility on our Soldiers.

I am confident that the vignettes offered in this volume will help leaders at all levels better understand and execute Mission Command.

David G. Perkins
Lieutenant General, US Army
Commanding

Introduction

In 2012, the US Army formally issued new doctrine on Mission Command, the philosophy and practice of command that serves as a foundation for Unified Land Operations. That doctrine defines Mission Command as, **"the exercise of authority and direction by the commander using mission orders to enable disciplined initiative within the commander's intent to empower agile and adaptive leaders in the conduct of unified land operations."** To provide a framework for the practice of Mission Command, the doctrine established six principles:

Build cohesive teams through mutual trust

Create shared understanding

Provide a clear commander's intent

Exercise disciplined initiative

Use mission orders

Accept prudent risk

Soon thereafter the Army began a focused effort to educate and train leaders on Mission Command as a way to prepare them for unpredictable and complex conflicts yet to appear on the horizon.

In early 2013, the Combat Studies Institute became involved in this effort by writing a series of Mission Command case studies for use at the US Army's Joint Readiness Training Center (JRTC) at Fort Polk. Those case studies form the core of this collection. Each case includes a brief account of a military action followed by an explanatory section that demonstrates how the case illustrates Mission Command principles. This structure was designed for use in training and schools but is equally conducive for self-study programs.

None of the 16 cases in this volume offer examples of leaders practicing Mission Command perfectly. Some of the actions described, in fact, come from early periods in which the lack of radio and other modern communications made any level of command and control very difficult to attain.

The real value of these cases lies in their ability to clearly convey how past leaders employed principles such as the use of commander's intent and the exercise of disciplined initiative to seize, retain, and exploit the initiative. In this way, the past breathes life into current doctrine, making it more tangible and understandable.

We at the Combat Studies Institute hope that these cases will enhance the way in which today's Soldiers understand the philosophy and practice of Mission Command as they prepare for future operations.

CSI-The Past is Prologue!

                              Roderick M. Cox
                              Colonel, US Army
                              Director, Combat Studies Institute

Contents

Section 1: Cases at Corps/Division Level

1. Failure of Command at Pea Ridge, 1862 1
2. Extending the Line at Little Round Top, July 1863 17

Section 2: Cases at Brigade/Regiment/Battalion Level

3. Nelson, Mission Command, and The Battle of Nile 31
4. Assault on Queenston Heights, October 1812 41
5. A Motorized Infantry Regiment Crosses the Meuse River, May 1940 53
6. Corregidor: Triumph in the Philippines 67
7. Assault River Crossing at Nijmegen, 1944 79
8. Sicily, 1943: Initiative Prevails at Biazza Ridge 89
9. Thunder Run in Baghdad, 2003 105
10. The Drive to Bastogne 119

Section 3: Cases at Company/Platoon/Squad Level

11. An Engineer Assault Team Crosses the Meuse, May 1940 133
12. Capturing Eben-Emael: the Key to the Low Countries 143
13. The Bridge at Mayenne, France 1944 155
14. The Victory at Tarin Kowt 165
15. The Attack on the Ranch House, August 2007 175
16. Operation NASHVILLE: Breaking the Taliban's Stranglehold in Kandahar, 2010 195

About the Contributors 205

List of Contributors

Richard V. Barbuto, Ph.D.

Anthony E. Carlson, Ph.D.

Mark T. Gerges, Ph.D.

Kendall D. Gott

Colonel Thomas E. Hanson, Ph.D.

Gregory S. Hospodor, Ph.D.

Kevin M. Hymel

John T. Kuehn, Ph.D.

John J. McGrath

Nicholas A. Murray, Ph.D.

Donald P. Wright, Ph. D.

Section 1: Cases at Corps/Division Level

Failure of Command at Pea Ridge, 1862
Colonel Thomas E. Hanson, Ph.D.

After a tumultuous summer of 1861, the state of Missouri remained a flashpoint in the Civil War west of the Mississippi. Confederate forces, unable to eradicate Union authority, retained control of the southern third of the state as the Lincoln Administration implemented the final phase of the Anaconda plan: the seizure of the Mississippi River. In late 1861, GEN Ulysses S. Grant designed a campaign to seize control of the upper Tennessee River early the following year. Doing so would require secure lines of communication from St. Louis and Cairo, Illinois, to keep his army supplied. After seizing Forts Henry and Donelson in February, Grant proposed to move deeper into west Tennessee to confront a rebel Army led by Confederate LTG Albert Sidney Johnston. Doing so, however, exposed Grant to attacks by Confederate forces in Arkansas. Johnston hoped to call on the 8,700 Confederate regulars in Arkansas to help block Grant's penetration. Viewing the situation from the strategic perspective, Confederate President Jefferson Davis and his cabinet hoped that continued military operations in Missouri would either cut Grant off from his base or prevent his campaign altogether. Union commanders, however, refused to cede the initiative to the Confederates. The resulting campaign ended with a decisive Confederate defeat at Pea Ridge (Elkhorn Tavern), Arkansas, forever ending Southern hopes of adding Missouri as a full member of the Confederacy while facilitating Union occupation of much of Tennessee.

Earl Van Dorn.
Courtesy Wilsons Creek National Battlefield
WICR 31608.

Although initially caught off guard by an unexpected winter campaign, Confederate forces owed their defeat at Pea Ridge, Arkansas, in March of 1862 to the limitations of and mistakes made by their commander, MG Earl Van Dorn. The federal commander, BG Samuel R. Curtis, better understood the requirements of both the art and science of mission command than Van Dorn, who ignored both the advice of his subordinates and his own senses in pressing an attack after the original goals became unattainable.

Samuel Ryan Curtis. Courtesy Wilsons Creek National Battlefield WICR 31443.

President Abraham Lincoln directed MG Henry Halleck, commander of the Department of the Missouri, to both keep Missouri for the Union and support military operations to defeat the rebellion. Halleck, in turn, organized the forces under his command to secure both objectives. Grant received command of two divisions of 17,000 soldiers and 13 gunboats to execute offensive operations. Curtis assumed command of the Army of the Southwest, a force of approximately 12,000 men that included a significant artillery capability. Curtis believed that to secure Missouri (and Grant's right flank) he needed to seize the initiative by conducting a winter campaign against Confederate forces led by former Missouri governor and would-be war hero MG Sterling Price.

Price's impressively-named Missouri State Guard was in fact a loosely-organized militia that numbered between 6,000 and 8,000 who were ill-equipped for active military operations. As many as 2,000 of these men lacked muskets. One contemporary described Price's force as "a mere gathering of brave but undisciplined troops, coming and going at pleasure." Nevertheless, as long as they remained in the field they posed a threat to Union control of Missouri.

The charismatic Price, a Mexican War veteran nicknamed "Old Pap" by his men, established a name for himself in August, 1861 when he and BG Benjamin McCulloch defeated Union forces under BG Nathaniel Lyon at Wilson's Creek, Missouri. Price remained in the southwest of Missouri throughout 1861 and into the new year while McCulloch returned to northwest Arkansas. Curtis, determined to rid Missouri of Confederate influence once and for all, established a supply base at a railhead in Rolla under the supervision of a promising young captain named Phillip H. Sheridan. Curtis then directed his men to shed much of their baggage and campaign gear so they could travel quickly. The Army of the Southwest set off from Lebanon, Missiouri on February 10, 1862. The weather became an adversary to both armies as temperatures plummeted below freezing and drifting snow blanketed the land.

Unprepared for battle, Price sent increasingly frantic pleas for help to BG James McIntosh, McCulloch's deputy. Rebuffed by McIntosh, Price abandoned Springfield, Missouri, and began a hasty retreat south, intent on joining with McCulloch's forces in northwest Arkansas. This was a bitter pill for Price, as he regarded McCulloch as a rival and his personal nemesis. President Davis and Confederate Secretary of War Judah Benjamin blamed Price for the discord; in response to Price's repeated plea to be commissioned a major general in the Confederate Army (and thus to outrank McCulloch), President Davis instead sent West Point graduate Van Dorn to assume command of the Confederate District of the Trans-Mississippi.

A veteran of the Eastern Theater, Van Dorn was a native of Port Gibson, Mississippi and thus a near neighbor to the Jefferson Davis family. Service in Mexico awakened a burning ambition in the dashing cavalryman that the outbreak of war in 1861 turned into an all-consuming passion for glory. Professionally, Van Dorn was a dim bulb; Confederate GEN Richard Ewell once remarked that Van Dorn had "learned all about commanding fifty United States dragoons and forgotten everything else" during his pre-1861 career. Historian Earl Hess observed that Van Dorn's "zeal for closing with the enemy was matched by his impatience with reconnaissance, logistics, and staff work of any kind." Van Dorn assumed command at Little Rock on 29 January 1862 and immediately began planning an invasion of Missouri. In his own words, the endstate he visualized was "to make a reputation and serve my country conspicuously or to fail. I must not, shall not, do the latter. I must have St. Louis—then Huzza!" Van Dorn hoped a spectacular victory would facilitate his triumphant return to the Eastern Theater, where the Confederate cavalry that he had trained was now garnering laurels under MG J.E.B. Stuart and not Earl Van Dorn.

Curtis' pursuit of Price forced Van Dorn to postpone his invasion of Missouri to defend Confederate soil. Price begged McCulloch for reinforcements but McCulloch refused. Angry at being ejected from Missouri and by McCulloch's snub, Price failed to notify McCulloch or Van Dorn of Curtis' presence until after he crossed into Arkansas on 16 February. The report sent the Confederates into a frenzy of activity as they hastened to assemble their forces. Van Dorn, who had yet to visit McCulloch or inspect his troops, began a nine-day overland trip from Pocahontas to Van Buren. Along the way, he fell into the icy Arkansas River and developed a serious fever, significantly degrading his already overtaxed intellect.

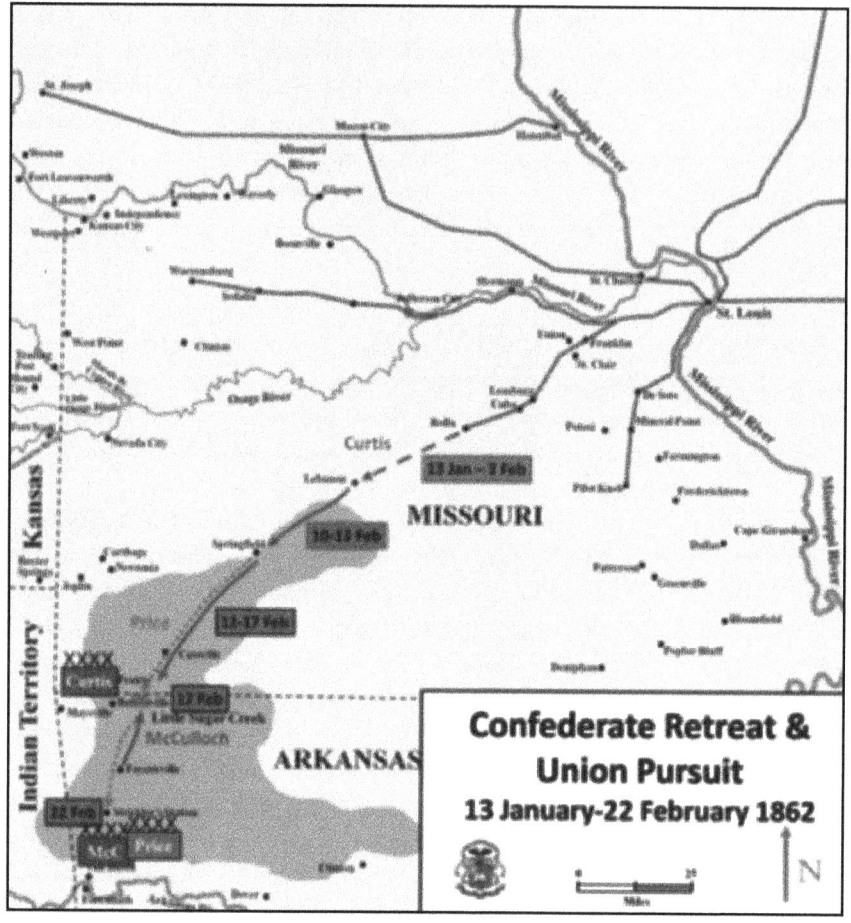

Figure 1.

Van Dorn's opponent was a complete opposite in temperament and training. Although both Curtis and Van Dorn were graduates of the United States Military Academy (1831 and 1842 respectively), all similarity ended there. Curtis almost immediately resigned his commission and practiced law in Ohio and later Iowa until elected to Congress in 1856. Volunteering to serve during the Mexican War, he saw no combat but served as a military governor of several conquered cities before returning to civil life. An early member of the Republican Party, he was considered for cabinet positions in the Lincoln Administration before resigning in favor of military service in 1861. Cool and thoughtful, with a thorough legal education and long life experience, Curtis possessed an agile mind and at Pea Ridge demonstrated an intuitive feel for events that served him well. He was also a shrewd judge of character who always looked to employ his subordinates at tasks for which they were best suited.

Like Van Dorn's force, Curtis' army included two very different sets of soldiers. Two of his small divisions were composed of a mix of native-born and immigrant troops from Illinois, Indiana, and Iowa. These elements were commanded by COL Jefferson C. Davis (3d Division) and COL Eugene A. Carr (4th Division). Like Curtis, they were graduates of the Academy and veterans of the Mexican War. Both men were also present for the defeat at Wilson's Creek the previous summer, a disaster for which many in the north blamed MG Franz Sigel, now Curtis' titular deputy and the driving force behind the other two divisions in the Army of the Southwest. Sigel, a German immigrant, graduated from the Military Academy at Karlsruhe and led anti-Prussian revolutionary forces in the revolutions of 1848 before coming to America. Settling in St. Louis, he became a leading pro-Union figure among German émigrés living in the United States. Commissioned a brigadier general of volunteers in 1861, Sigel led one contingent of federal troops at Wilson's Creek while BG Nathaniel Lyon led the other. Sigel's half-hearted attempt to link up with Lyon doomed the latter to defeat (and death) while Sigel escaped safely amid whispered charges of cowardice. Sigel's influence with German-Americans, however, prevented Lincoln from removing him. In early 1862 Sigel's two-division force was composed almost entirely of German emigrants from Missouri and Illinois, commanded by COL Peter J. Osterhaus (1st Division) and BG Alexander S. Asboth (2d Division). Both had been officers in European armies before coming to the United States.

Crossing into Arkansas on 17 February 1862, Curtis' lead elements caught up with Price just as the latter's exhausted militia linked up with McCulloch's Confederate regulars. A sharp engagement took place about four miles south of Elkhorn Tavern on the Wire Road (so named because it followed the telegraph wire route) causing Curtis to halt his pursuit and consolidate his force. The Confederates withdrew south to Fayetteville, which they abandoned a few days later after plundering the town and leaving their fellow citizens unprepared for the remaining months of winter. Curtis now faced an operational dilemma. He had technically exceeded his authority by invading the Confederacy. Halleck had ordered him to eliminate the threat posed by Price. Driving Price into the arms of another rebel army had not necessarily done that. As long as Price's army existed it could return to Missouri. Unlike most of his contemporaries, Curtis clearly understood that his mission required the destruction of Price's army, not simply its removal from Union territory. However, he was now over 200 miles from the railhead at Rolla and dependent on wagon trains to resupply his force. In addition, he had depleted his force by nearly 20% in order to garrison Springfield, Missouri and other critical locations along his line of march. Finally, the Ozark Plateau in late winter offered little food for man or horse. Curtis realized he could not pursue Price and McCulloch deeper into Arkansas without risking being cut off from his supplies but he could not withdraw and cede the operational initiative to Van Dorn. Therefore, he resolved to defend Missouri from just inside the Arkansas state line. Curtis established his defenses astride the Wire Road east of Bentonville, spreading his forces to facilitate foraging. From 19 February to 5 March, Sigel's two divisions occupied positions west of the Wire Road near Bentonville while Carr's and Davis' divisions camped near Curtis' headquarters at Cross Hollow on the Wire Road. Carr dispatched a 700-man detachment of infantry, cavalry, and artillery to Huntsville, 35 miles southeast of Cross Hollow, on 4 March in an attempt to arrest local Confederate leaders. Finding nothing significant, COL William Vandever decided to spend the night of 5-6 March in Huntsville and return during daylight hours.

Earl Van Dorn, exhausted and feverish after his overland trek across Arkansas, met with Price and McCulloch at Strickler's Station on 3 March. Van Dorn's adjutant, Dabney Maury, immediately noted the difference between Van Dorn's chief subordinates. Price had hosted Van Dorn to a sumptuous feast upon the latter's arrival the night prior despite the meager rations issued available for Price's men. McCulloch's headquarters, however, was spartan in furnishings and businesslike in atmosphere. Maury recognized that unlike Price, McCulloch and his staff possessed "the stern seriousness of soldiers trained to arms." Moreover, McCulloch presented Van Dorn with a detailed description of Curtis' dispositions and a proposed plan of attack, for which he had issued preparatory guidance two days earlier. Van Dorn instantly grasped that Curtis' two wings were not mutually supporting, and "resolved to attack him at once," believing that if he smashed the Army of the Southwest he could still get to St. Louis and glory. He ordered an attack for the following day, 5 March, 1862.

Van Dorn's plan was elegantly simple. McIntosh's brigade of 3,000 Texas and Arkansas cavalrymen would march north on the Wire Road and demonstrate in front of Curtis' position at Cross Hollow. By doing so they would screen the remainder of McCulloch's division (5,700-soldiers with 18 artillery pieces) and Price's division of about 6,800 Missourians with 47 artillery pieces. Van Dorn believed his infantry could move north and defeat Sigel before he or Curtis could identify the threat because Curtis possessed a significantly weaker cavalry capability. To ensure rapid move-

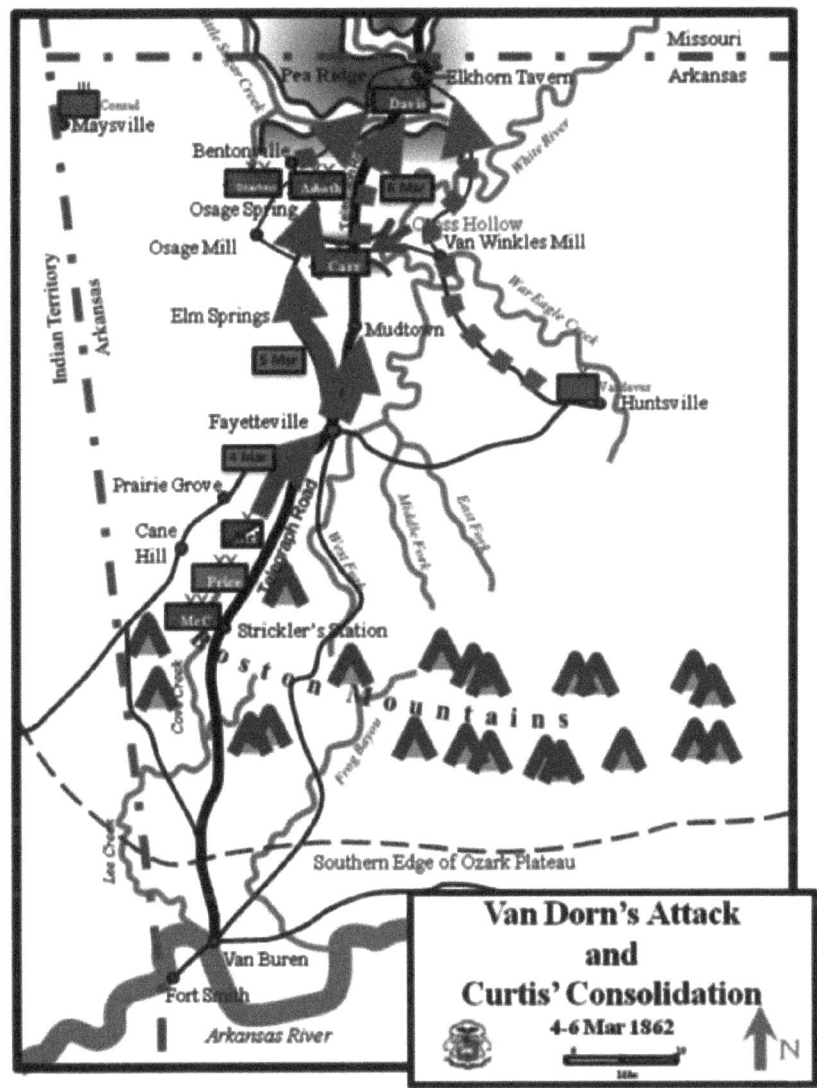

Figure 2.

ment, Van Dorn issued explicit orders to travel lightly. Each soldier would carry only his rifle, forty rounds of ammunition, a single blanket, and rations for three days. Only an emergency supply of ammunition and a single day's rations would accompany the attacking column. Van Dorn planned for his men to resupply themselves from captured stores after defeating Curtis. He gave no thought to other contingencies or the fact that after defeating the Federal Army he would be responsible for feeding prisoners as well as his own soldiers.

Van Dorn made several mistakes at this crucial juncture. First, he did not know his subordinates except by reputation. Since neither Price nor McCulloch were graduates of the Military Academy, President Davis and Secretary Benjamin considered them equally unqualified. Van Dorn shared these prejudices and did not bother to learn about either man's skills or experience. Had he done so, he would have discovered that McCulloch was a veteran of both the Texas Revolution and the Mexican War and served as a Texas Ranger for two decades. He possessed unsurpassed battlefield experience against both conventional and unconventional opponents. Universally admired by his troops, McCulloch was a no-nonsense commander and able tactician. His thorough preparations enabled Van Dorn to attack as soon as he arrived in northwest Arkansas.

Second, Van Dorn's indifference to the condition of his men bordered on dereliction of duty. Price's men had walked over two hundred miles in less than ten days with little food. McCulloch's troops had shared their food with Price's men, depleting their own supplies and then had burned tons of food and equipment when they abandoned Fayetteville. Living under canvas in the snow and sleet of the Boston Mountains in February and March had not allowed Price's men to recover nor did it allow McCulloch's men to retain their strength. Nevertheless, Van Dorn's plan depended on his soldiers' ability to ignore fatigue, hunger, and harsh environmental conditions to rapidly close with Sigel's stationary divisions.

Third, Van Dorn's quest for quantitative superiority diluted the quality of his force when he accepted two late additions to what he now called his "Army of the West." The first was two green (and unarmed) regiments of Arkansas infantry. These men were ultimately left behind but their presence further burdened an already deficient supply of food and equipment. The second addition was two small regiments of Indian "volunteer" infantry (1st and 2d Cherokee Mounted Rifles, about 700 men altogether) and two companies of Texas cavalry. This unsavory group was commanded by BG Albert Pike, a morbidly obese political appointee who combined poor judgment with a complete lack of qualification for his position. Pike's Cherokees were a mixed bag of adventurers, restless youth, and brigands, most of who agreed to serve only after pocketing Pike's generous bounties. Distrusted by the Confederates and hated by the Federals, their performance on and off the battlefield left much to be desired.

Van Dorn's army broke camp on 4 March and began moving north into a blizzard. The Confederates' enthusiasm lessened with each step as Van Dorn set a hellish pace while riding in his covered ambulance. One of

Price's men remarked afterward that Van Dorn "had forgotten he was riding and we were walking." The column halted for the night in Fayetteville. The Confederates spent a cold night wrapped in their single blankets in the burned-out ruins of the town. The next day's march was equally miserable with snow and freezing temperatures combining with the stress of the march to cause numerous stragglers. Late in the day, the 3d Texas Cavalry blundered into a Union outpost on the Elms Springs Road south of Bentonville. The Federal infantrymen defended their position before withdrawing in good order to report what they had seen. Van Dorn's hope for tactical surprise disappeared with them.

Before news of Van Dorn's movement reached Curtis, he had already decided to shrink his footprint by concentrating on the high ground just north of where the Wire Road crossed Little Sugar Creek. Multiple reports of a general Confederate movement reinforced his intuition and Curtis lost no time in ordering Sigel to bring his divisions east. Meanwhile, COL Vandever and his little detachment set out to return to the main body at 0300 on 6 March. Spurred by reports of approaching Confederate cavalry, Vandever's column closed on the new defensive positions at 2000 after covering the 35 miles without loss of a single soldier, horse, or artillery piece. By then, Sigel had begun moving his forces, and after personally commanding the rear guard in several hot engagements, he joined Curtis two hours after Vandever. Van Dorn's plan lay in ruins. Not only had Curtis and Sigel not been caught unaware, the Confederates now faced a united Federal Army entrenched on commanding high ground.

At this point, had Van Dorn chosen to make effective use of McIntosh's cavalry brigade, he could have fixed Curtis in his positions with superior numbers of infantry and a large artillery contingent while the cavalry severed Curtis's single line of communication. Doing so would have required considerable patience which Van Dorn did not possess. Instead, at a council with McCulloch and Price, a feverish Van Dorn accepted the former's suggestion that the entire Army attempt a turning movement by sidestepping the Federal position to the northwest, using the Bentonville Detour to travel around Pea Ridge and seize the Detour's junction with the Wire Road near Elkhorn Tavern. Worse, Van Dorn demanded that the necessary movement begin immediately on the night of 6 March. His men were exhausted, cold, and hungry with empty ration bags owed to Van Dorn's strict orders travel lightly. McCulloch asked Van Dorn to reconsider his movement order to allow the soldiers to sleep for a few hours and attack the following day. Van Dorn would hear none of this. His glory depended on beating Curtis to the punch and forcing him to surrender.

Van Dorn wrongly assumed that Curtis was concentrating in order to retreat, not fight, and closed his mind to any alternative. Curtis' subordinates, however, knew their commander sought to destroy the Confederate force and sought opportunities to set the conditions for success. COL Grenville Dodge, commanding a brigade in Carr's division, suggested to Curtis that the Bentonville Detour be sown with obstacles to hinder exactly what Van Dorn now proposed to do. As a result, when the Confederates moved into the narrow defile, they soon encountered two separate "mazes" of felled trees which Dodge's men constructed. Moreover, poor Confederate staff work placed the infantry and artillery ahead of the cavalry in the movement order. As a result, the rebels' first indication of a problem was when Price's infantry blundered into the fallen trees just after midnight. Instead of being formed for battle as the sun rose, Van Dorn's column was stretched out over a dozen miles and men by the hundreds straggled in the woods along the road prostrated by hunger and fatigue.

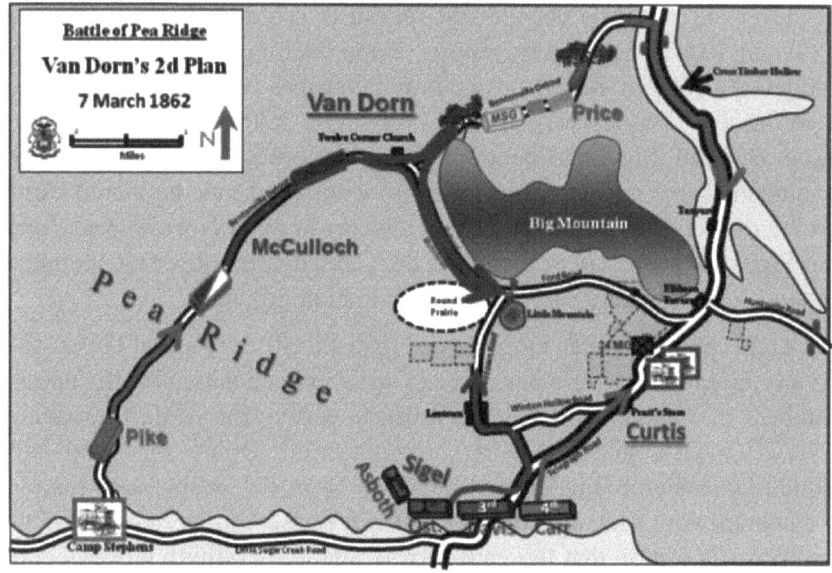

Figure 3.

Like his initial plan, Van Dorn's second attack order seemed simple. Moreover, the intended objective of the attack was a high payoff target. The entire supply train of the Army of the Southwest was scattered across the open ground adjacent to the Elkhorn Tavern. Had the Confederates moved successfully to the Federal rear and seized the Wire Road at the tavern, Curtis would have been faced with re-orienting his force 180 degrees and launching a hasty attack, or surrender. Because Price's exhausted infantry could not concentrate, however, by the time Van Dorn began his attack, Carr had moved the Union trains out of danger and had established a defensive line against which Price's men would batter themselves for most of the afternoon. In an attempt to remain on schedule, Van Dorn had modified his plan when he realized he could not bring his entire force to bear in time for battle. Realizing the Detour was hopelessly clogged, Van Dorn directed McCulloch to leave the Detour and take the Ford Road laterally across Pea Ridge and link up with Van Dorn's force at Elkhorn Tavern. In doing so, however, McCulloch marched blindly into a decisive engagement of his own, leaving Price to attack unsupported.

McCulloch welcomed his new orders after spending most of the evening immobile in subfreezing temperatures due to the congestion on the Bentonville Detour. Extricating his men, his entire division moved east toward the pass between Big Mountain and Little Mountain. Even before sunrise, however, federal pickets identified the threat and Curtis dispatched Osterhaus with instructions to locate and engage this force. Osterhaus deployed three infantry regiments with three batteries and several cavalry companies into Oberson's Field, facing north. A spoiling attack by the 3d Iowa Cavalry met with disaster when nearly all of McIntosh's cavalry brigade charged their attackers. Three guns were lost and over a hundred men were killed or captured. In their only active involvement in the battle, Pike's Cherokees moved in behind the rebel cavalry and several Iowans were scalped and mutilated before the Indians were frightened away by Union artillery fire.

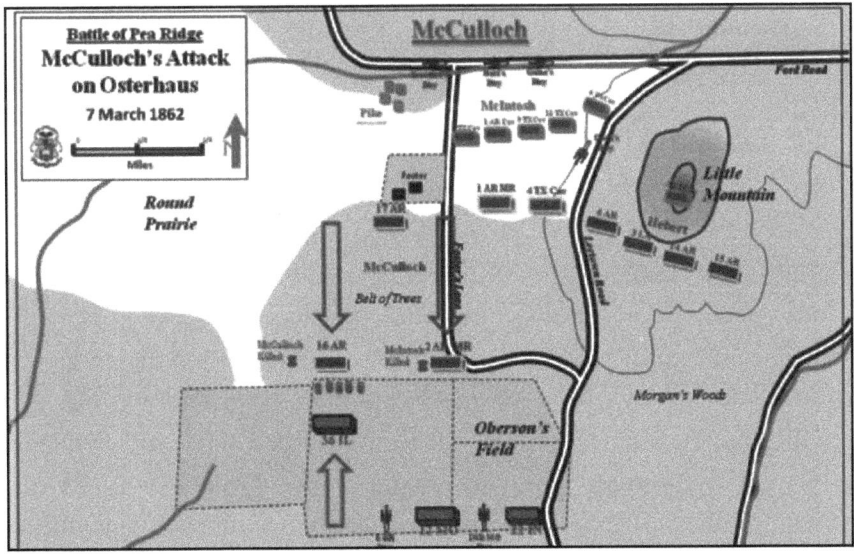

Figure 4.

Rather than bypass Osterhaus, McCulloch halted his force in column on the Ford Road. McCulloch assumed command of half of the infantry brigade himself, leaving COL Louis Hébert in command of the other half. He ordered Hébert to form for a general attack and move forward when he heard McCulloch's wing engage the Union line. McIntosh was directed to reform his cavalry and prepare to resume the march. Only Goode's Texas battery was unlimbered to support the assault while three others remained on the road. Union 12-pound howitzers, firing blindly, inflicted severe losses on the infantry ranks and their shots went mostly unanswered. Chance again intervened on the battlefield when McCulloch was shot down while making a personal reconnaissance. His death was kept secret for fear of demoralizing his men. McIntosh was informed and assumed command but he was killed in exactly the same way less than two hours later. Hébert mistook the fusillade which killed McIntosh as the signal to advance. His well-trained Louisianans, the cream of Van Dorn's army, were mowed down when two additional Union brigades arrived to reinforce Osterhaus and caught Hébert's men in an L-shaped kill sack. Hébert himself was captured after wandering through the Union lines delirious from thirst, hunger and fatigue. What remained of McCulloch's command disintegrated. Some retreated west to the Confederate trains while the rest, under Pike, backtracked to the Bentonville Detour and followed the Confederate main body. They did not link up with Van Dorn until dawn of the following day.

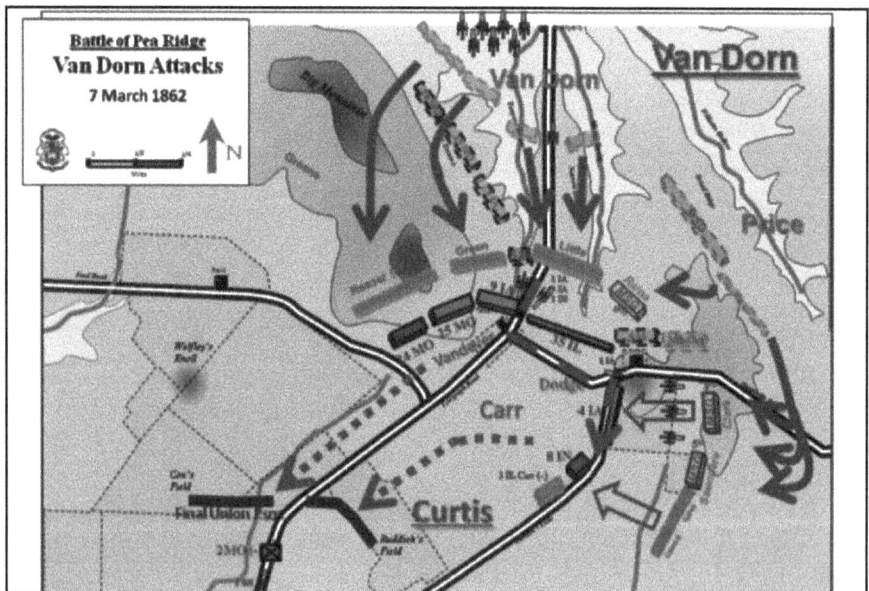

Figure 5.

When he sent Osterhaus to the west, Curtis was also aware of Price's movement on the Bentonville Detour. What he did not yet know was which of the three threats (south, north, or west) was Van Dorn's main effort and which were feints. On his own volition, COL Dodge directed his entire brigade to abandon their positions overlooking Little Sugar Creek and move to Elkhorn Tavern, so sure was Dodge that the main Confederate attack would come in that area. Coming upon Dodge's troops, Curtis ordered Dodge (who was Osterhaus' subordinate) to reinforce COL Carr's weak division at the Tavern around 1100. Shortly afterward, Curtis rode to his forward positions overlooking the creek and after listening to the lackluster firing of a diversionary Confederate force, directed Asboth to abandon the entrenchments above Little Sugar Creek and form a defensive line about 1000 yards west of Elkhorn Tavern. By the time Price's exhausted Missourians had fought their way out of Cross Timber Hollow and onto Pea Ridge, they were incapable of overcoming the Federal defenses. They passed the night of 7-8 March where the advance had stopped, not bothering to entrench.

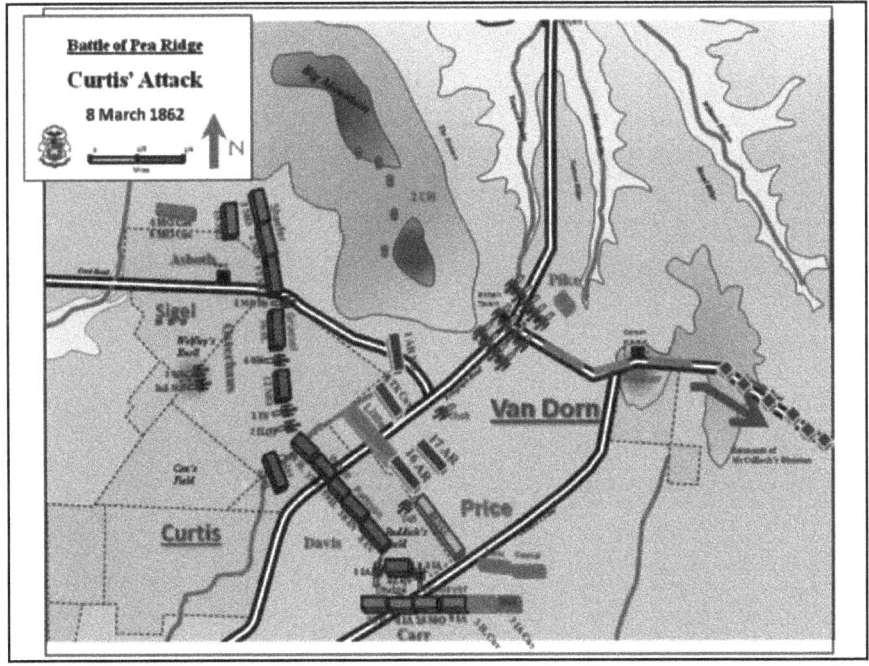

Figure 6.

All through the night, while Pike led the remnants of McCulloch's division to Van Dorn, the Union leadership rearranged their defenses. By sunrise on the 8th, Curtis had successfully reoriented his entire army to face Van Dorn. In the growing glare of a clear frozen day, exhausted rebels awoke to see the sun glinting on the bayonets and illuminating the colors of the Army of the Southwest, arrayed across their front and supported by nearly 50 guns. MG Sigel, after playing almost no part in the previous day's fighting, personally sited the Union artillery. Van Dorn's artillery, by contrast, had used up almost all its ammunition during the previous day's attacks and again poor staff work prevented timely resupply. Out of ammunition, rebel gunners resorted to loading nails, horseshoes, tools, and forks and knives but could not prevent the surging lines of blue from prevailing. In the only instance of its kind in the Civil War, the entire Army of the Southwest charged Van Dorn's lines, sweeping the exhausted Confederates into a rout so complete that Missouri and Arkansas ceased to be active theaters of war for over a year. Although not destroyed, Van Dorn's army required months to reconstitute and therefore could not reinforce Johnston for his battle against Grant at Shiloh a month later.

For further reading:

William L. Shea and Earl J. Hess. *Pea Ridge: Civil War Campaign in the West*. Chapel Hill: University of North Carolina Press, 1992.

Earl J. Hess, Richard W. Hatcher III, William Garrett Piston, and William L. Shea. *Wilson's Creek, Pea Ridge, & Prairie Grove: A Battlefield Guide with a Section on the Wire Road*. Lincoln, NE: University of Nebraska Press, 2006.

William L. Shea. *The Campaign for Pea Ridge*. National Park Civil War Series. Washington, DC: Eastern National Publishing, 2001.

The Six Principles of Mission Command

1. Build Cohesive Teams through Mutual Trust
2. Create Shared Understanding
3. Provide a Clear Commander's Intent
4. Exercise Disciplined Initiative
5. Use Mission Orders
6. Accept Prudent Risk

Mission Command in the Pea Ridge case

1. Build Cohesive Teams through Mutual Trust. Van Dorn commanded through force of will rather than by understanding his subordinates and seeking to secure their wholehearted support. His lack of understanding of BG McCulloch's experience and capability was particularly damaging during the campaign and when the battle began.

2. Create Shared Understanding. Van Dorn relied on McCulloch and Price to provide him with the details of terrain and the Union dispositions. He made no plan to utilize his vastly superior cavalry capability as a reconnaissance or raiding force. As a result, he allowed Curtis to reorient his forces without distraction and condemned his men to march along the Bentonville Detour into the obstacles created by Dodge's men. A quick reconnaissance by cavalry would have revealed the blockage, and enabled the entire force to shift to the Ford Road. Had Van Dorn chosen that course of action he would have overwhelmed Curtis.

3. Provide a Clear Commander's Intent. Van Dorn's intent, "personal glory," was hard to measure, and the key tasks provided to his subordinates were not always clearly explained. As a result, McIntosh allowed himself to be surprised by the Union outpost on Elm Springs Road because he failed to understand the criticality of the counter-reconnaissance fight.

4. Exercise Disciplined Initiative. No one in Van Dorn's army appears to have understood the importance of flexibility, or of the requirement to adapt to changing conditions by altering plans while staying within the commander's intent. McCulloch allowed himself to be sucked into a meaningless engagement rather than come to Price's support on 7 March. As a result, Price was unable to overcome Union resistance. Similarly, McIntosh, Hebert, and Pike demonstrated only a slavish obedience to the plan. McIntosh and Hebert in particular should be faulted for not pressing McCulloch to either bypass Osterhaus or apply the full weight of the division's artillery.

5. Use Mission Orders. Van Dorn did issue mission orders throughout the short campaign, but only because he preferred to avoid details.

6. Accept Prudent Risk. Van Dorn accepted risk in his moving to Curtis' rear, where he was also putting Curtis between himself and his own line of communication. Had he been able to concentrate and attack on schedule, it would have paid off. But by ignoring the condition of his men, the weather, the terrain, and the fact that he failed to clearly communicate to McCulloch not to delay his movement on the Ford Road, he himself set the conditions for his own defeat.

Extending the Line at Little Round Top, July 1863
John J. McGrath

At the Battle of Chancellorsville in May 1863, Union forces failed to dislodge GEN Robert E. Lee's Confederates from positions south of the Rappahannock River in Virginia halfway between Washington and Richmond. In response to his victory, Lee led his forces on an invasion of the northern states of Maryland and Pennsylvania. This invasion set the stage for an ultimate showdown between his forces and those of the Union Army of the Potomac commanded by MG Joseph Hooker. When a chance encounter between Confederate infantry and Union cavalry northwest of the crossroads town of Gettysburg on 1 July 1863 evolved accidentally into a major battle, Lee had his showdown.

As most of the Confederate forces were closer to Gettysburg than were the Union forces, they were able to consolidate more quickly north of Gettysburg and overwhelm the Federal troops, forcing them to retreat through the town and occupy high ground to the south as darkness arrived on the battlefield. Several days earlier, MG George Meade, formerly Fifth Corps Commander, had replaced Hooker as the commander of the Army of the Potomac. During the night, the bulk of both armies arrived. Meade decided to defend the high ground south of Gettysburg the next day while awaiting the arrival of his last two (Fifth and Sixth) Corps. At the same time, Lee decided to follow up his success with an attack against Meade's forces on 2 July 1863.

Lee decided to have his Second Corps, commanded by LTG James Longstreet, mass against the Union left (southern) flank and attack it in a maneuver designed to overwhelm the Union position by outflanking it. Lee and Longstreet had used just such a maneuver successfully almost a year earlier at the Second Battle of Bull Run. Longstreet intended to march his command using covering terrain in order to surprise the Federal defenders and then to attack *en echelon*, a technique in which each of his units would attack in succession from south to north in order to have maximum effect of the defenders, who could be expected to be responding to the previous attacks when the subsequent ones commenced.

Figure 1. Map. General Situation, 1 July 1863.

Meade planned on defending the high ground south of the town with a continuous line anchored on two hills (Culp's Hill and Little Round Top) on each flank. The position from north to south consisted of Culp's Hill, Cemetery Hill, Cemetery Ridge, and Little Round Top and formed a semi-circle around Culp's and Cemetery Hill with a longer line extending south along Cemetery Ridge to Little Round Top. Initially defending the extreme southern end of this line was MG Daniel Sickles' Third Corps. Sickles was still smarting from being ordered to retreat from a strong position at Hazel Grove during the Battle of Chancellorsville several weeks earlier and being forced to fight on less favorable ground. Accordingly, using his own initiative, he advanced his corps forward to the west to positions along the

Emmitsburg Road which he felt were superior to those assigned him. To cover this advanced position, Sickles placed two brigades on his left flank. However, Little Round Top, to the rear of the new Third Corps line, was left unoccupied.

Sickles initiative failed to take into account the intent of his commander, Meade, for fighting the battle. Meade planned to fight a defensive battle along a continuous line that did not have any gaps and was anchored by two prominent terrain features that would make it difficult for the Confederates to outflank it. Sickles' deployment, conducted in a manner uncoordinated with the rest of the army, left Meade's left flank exposed right at the very place where Lee planned to attack with Longstreet's corps.

Little Round Top, while smaller in elevation than its southern neighbor, Round Top, was of greater military significance than the higher hill mass, both of which were covered by extensive foliage, because local farmers had recently cleared the western slope of Little Round Top of cover giving troops located on the hilltop extensive fields of observation and fire to the west towards the Emmitsburg Road and Seminary Ridge (where the bulk of the Confederate forces were assembled) as wells as to the north along Cemetery Ridge. A military force occupying Little Round Top would, therefore, possess a great advantage.

With Sickles occupying forward positions along the Emmitsburg Road, Longstreet's attack force, after a laborious and convoluted march behind Seminary Ridge that took most of the day of 2 July, emerged in covered attack positions on Sickles' left flank late in the afternoon. Longstreet intended to open his attack with MG John Hood's First Division followed by MG Lafayette McLaws' Second Division, to Hood's right opposite Sickles' main line. From their start positions, Hood's brigades would overlap and outflank the two brigades that Sickles had placed to cover his left and advance directly on Little Round Top. Unless Union forces were promptly moved to the hill, Hood threatened to outflank the whole Union position without much of a fight. Such a situation had doomed the Federal defenders at the Battle of Second Bull Run.

Meade had been holding the recently arrived Fifth Corps in reserve to the rear of the main position when his last major unit, the Sixth Corps, arrived. The Union commander decided to reinforce the left flank with the Fifth Corps, placing the tired Sixth Corps in reserve. Accordingly, Meade rode with his staff to the left in advance of the corps to determine the best location for the new unit and then discovered that Sickles had moved his troops forward of their assigned positions. At about the same time, Hood's men began their attack. Since Hood's right wing was close to Sickles' troops, they opened fire from a distance, indicating to Meade that it was too late to have Sickles return to his assigned posting. To accommodate the gap in the lines, Meade now ordered MG George Sykes' Fifth Corps, already on its way to the left flank, to reinforce Sickles. Meanwhile, Meade noticed the danger in the vicinity of Little Round Top and instructed his chief engineer, BG Gouverneur Warren who was accompanying him, according to one of Warren's aides, as follows, "Warren! I hear a little peppering going on in the direction of the little hill off yonder. I wish that you would ride over and if anything serious is going on … attend to it."

With this guidance, Warren and several aides rode to the summit of Little Round Top and found it occupied only by a small signal station. Not sure of the enemy situation, the chief engineer sent a courier down to a battery posted with Sickles' flank guard west of Little Round Top and had the battery fire a round over the dense foliage to the southwest of Little Round Top, vegetation which could cover the advance of a considerable body of enemy troops. As the round passed over the trees, Warren could see the gleam from the bayonets of the advancing Confederate infantry as the men flinched at the sound of the cannon shot. It was obvious that enemy forces were located in a position to outflank not only Sickles' corps but the entire army.

Warren knew he had to take immediate action. He promptly sent couriers to Sickles requesting a unit to garrison Little Round Top. Since the Third Corps was now in action, Warren felt that he had little hope of getting help from that quarter. However, he did notice that the first elements of Sykes' Fifth Corps were arriving along the road north of Little Round Top, sent by Meade to reinforce Sickles. Warren's courier to Sickles, 1LT Ranald Mackenzie, had received the expected negative response to his request for aid. However, while returning to Warren, Mackenzie ran into Sykes who agreed to send a brigade from his lead division to Little Round Top. When Mackenzie was unable to immediately find the division commander, the commander of the lead brigade, COL Strong Vincent, realized the gravity of the situation and volunteered to move his brigade without delay to Little Round Top. The brigade, officially the Third Brigade, First Division, Fifth Corps, consisted of four regiments of: 16th Michigan, the 44th New York, the 83d Pennsylvania, and the 20th Maine. Feeling the west side of the hill was protected by the troops of the Third Corps he could plainly see from the height, Vincent deployed his brigade on the southern side of Little Round Top. As deployed, the brigade had no troops on either its left or right flanks. The left flank, defended by the 16th

Michigan, was protected at least in the abstract, by the distant Third Corps but the left flank, where the 20th Maine was positioned, was the extreme left flank of the entire army, at least until reinforcements arrived. Within minutes of the brigade's deployment, it was attacked by two brigades of the Confederate right flanking force.

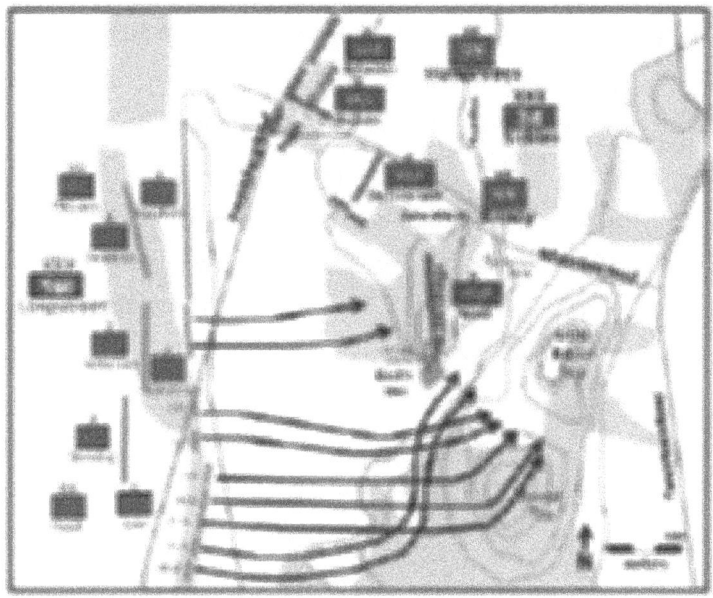

Figure 2. Map. Longstreet's Attack.

The two brigades were from Hood's division. On the extreme right, was BG Evander Law's brigade of five Alabama regiments. To Law's immediate left, advanced BG Jerome Robertson's brigade with two Texas regiments. The pair of brigades had conducted an exhaustive advance, covered by the wooded terrain. In the process, Law's two left regiments got tied down in the fighting with Sickles flank guard around a rocky outcropping dubbed Devil's Den. Robertson took up the space in the middle of Law's brigade. The deployment of Vincent's brigade placed it right in front of Law and Robertson's advance on Little Round Top. The tired Confederates twice advanced out of the cover of the trees and charged up the rugged, rocky, open terrain on the south side of Little Round Top. Vincent's men fought heroically and repulsed these two attacks. The Confederates prepared for a third advance.

Meanwhile, the Fifth Corps artillery chief had ordered a six-gun battery commanded by 1LT Charles Hazlett to move to Little Round Top to support the corps projected move to reinforce Sickles. It took some time for the guns to be manhandled to the top of the hill. Hazlett was familiar with Warren, having supported the brigade he commanded at the Second Battle of Bull Run the year before. Warren briefly helped to manhandle the guns up the slope. However, he soon rode off to find reinforcements himself.

Warren went down to the east-west road north of Little Round Top where he encountered BG Stephen Weed's brigade (Third Brigade, Second Division, Fifth Corps) moving westerly on the road, the advance element of Sykes' second division marching forward to reinforce Sickles. When Warren arrived, Weed, the brigade commander, was forward with Sickles. Leading the brigade's march was COL Patrick O'Rorke, the commander of the 140th New York.

COL Patrick O'Rorke
National Archives

Warren knew O'Rorke well. He had been an instructor at West Point when O'Rorke finished first in the class of 1861. Warren had previously commanded the same brigade during the Seven Days Battles, Second Bull Run, Antietam, and Fredericksburg. O'Rorke had commanded it at Chancellorsville. The 140th New York (under O'Rorke) had been part of the brigade since the Battle of Fredericksburg, when Warren still was in command. Of the other three regiments in the brigade, one (the 146th New York) had served under Warren and also contained former members of his original regiment (the 5th New York). The other two regiments (the 91st and 155th Pennsylvania) were veteran units but had previously served in other brigades.

Warren rode up to O'Rorke and asked for a regiment to man the hill. The New Yorker told his former commander that his current brigade commander (Weed) had moved forward and expected the brigade to follow. Warren told him, "Never mind that, bring your regiment up here and I will take responsibility." Given his knowledge of Warren's character and tactical abilities, O'Rorke did not hesitate further. He turned his regiment to the left and, led by one of Warren's aides, began climbing the hill.

Although their first two attacks had been repulsed, the Confederates still had hopes of breaking through the Federal defenses. On the right, the 15th Alabama had discovered the Union left flank and Law felt one more push could collapse it. On the left, Robertson's two regiments were now reinforced with Law's 48th Alabama, a unit which had been delayed by the fighting near Devil's Den but was now free to advance to the left of the two Texas regiments. The addition of the 48th Alabama meant that the Confederate attack frontage on Vincent's left opposite the 16th Michigan now overlapped that of the defenders. The 48th Alabama would be able to outflank and overwhelm the 16th Michigan while it was fighting to its front against Robertson's Texans. With other Confederate units attacking the rest of the brigade, Vincent's command was threatened with destruction. The battle soon became intense between the 16th Michigan and the 48th Alabama and 4th and 5th Texas. Vincent personally commanded until he was mortally wounded. It seemed his brigade was about to collapse.

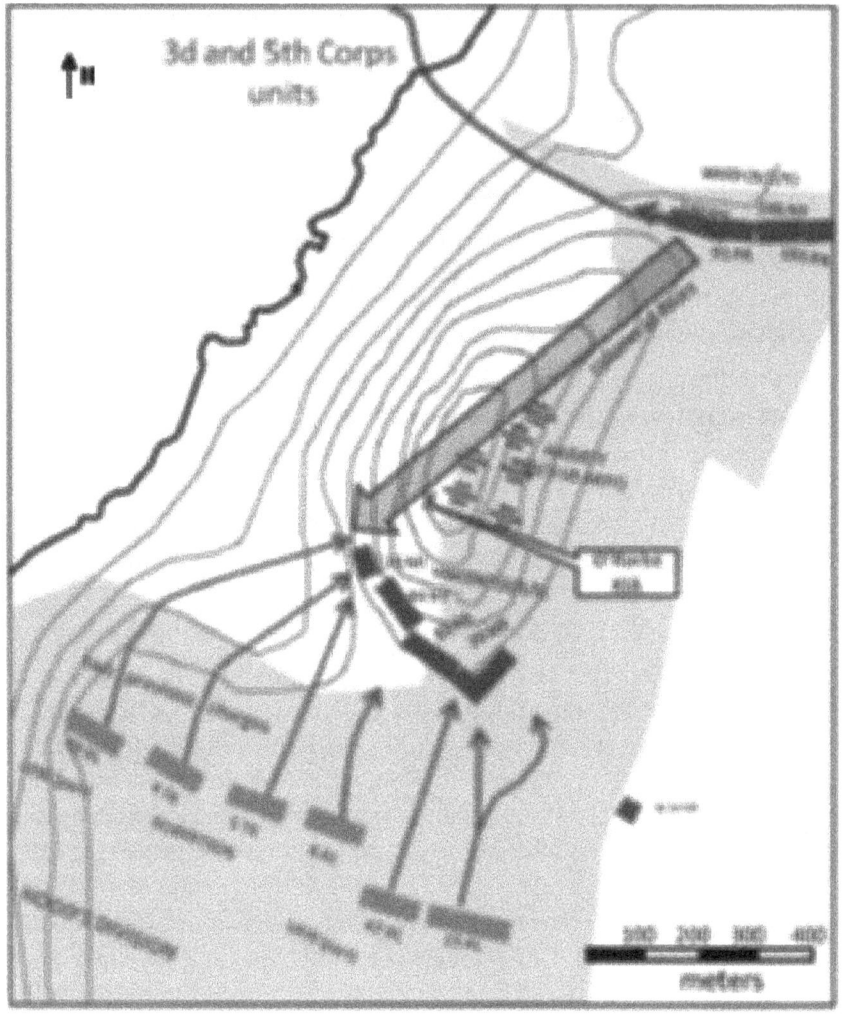

Figure 3. Map. O'Rorke Reinforces Little Round Top.

At this moment, O'Rorke's men appeared over the top of the hill advancing quickly in a column of fours around Hazlett's guns and aiming for the exact point of the Confederate breakthrough. O'Rorke, with sword drawn, led the regiment down towards the enemy. At this point, one of the advancing Alabamans fired at and killed O'Rorke. Rather than demoralizing the New Yorkers, O'Rorke's death made the men angry and more aggressive. The regiment opened fire and drove down the forward slope of the hill, setting up a defensive line to the right of the 16th Michigan. After minutes of fierce close combat, the Confederate attack was spent and the Southerners retreated to the cover of the trees at the bottom of the hill. Although there were still some further attacks on the front of the 20th Maine to the left, these were ultimately repulsed. The 140th New York had saved Vincent's men from being overwhelmed on the right and had secured Little Round Top for the rest of the battle. This short defensive battle had cost the regiment dearly. Out of about 500 men, casualties were 25 dead including COL O'Rorke, 89 wounded and 18 missing.

The regiment was soon reinforced by the rest of Weed's brigade. The brigade had continued down the road in response to its assigned mission and linked up with its commander. BG Stephen Weed had been recently promoted from an artillery battery commander to command of a brigade. As an artilleryman, Weed's last assignment was as commander of the Fifth Corps artillery brigade. One of Weed's aides was Warren's younger brother. While forward, a courier informed Weed that his Sykes, his corps commander, had changed the brigade's mission, the unit would join the 140th New York on Little Round Top. By the time Weed's men arrived on the hilltop and took up positions to the right of the 140th, the major action was over. However, Confederate snipers continued to fire at the defenders. Both Weed and Hazlett were mortally wounded by these sharpshooters.

The defense of Little Round Top succeeded for several reasons. Meade gave Warren guidance and expected that he would do what was necessary to protect the army's left flank. Warren, in turn, displayed vigorous initiative, energetically seeking out units to fill the gap on the hill. As a former brigade commander In the Fifth Corps, he was familiar to all the senior officers he encountered. This familiarity enhanced his ability to persuade commanders to move their troops according to his general instructions even though they were not under his command. O'Rorke, in particular, responded vigorously and, while he gave his life for this reaction, the arrival of his troops saved the day for the Union position just in time.

For Further Reading

Robert U. Johnson and Clarence C. Buel, eds. *Battles and Leaders of the Civil War*, 4 vols. New York: Century Company, 1884-89. Reprint, New York: Thomas Yoseloff, 1956.

Oliver Norton. *The Attack and Defense of Little Round Top, Gettysburg, July 2, 1863*. New York: Neale, 1913.

Harry Pfanz. *Gettysburg: The Second Day*. Chapel Hill, NC: University of North Carolina Press, 1987.

United States War Department. *The War of the Rebellion: a Compilation of the Official Records of the Union and Confederate Armies.* Washington, DC: War Department, 1880-1901), Volume 27, Part 1.

The Six Principles of Mission Command

1. Build Cohesive Teams through Mutual Trust
2. Create Shared Understanding
3. Provide a Clear Commander's Intent
4. Exercise Disciplined Initiative
5. Use Mission Orders
6. Accept Prudent Risk

Mission Command in the Little Round Top case

1. Build Cohesive Teams through Mutual Trust. All units of this operation were veteran organizations, most of which had worked together in the past. Warren had previously commanded Weed's brigade and knew all its senior officers. Weed had previously been an artillery battery and brigade commander who had supported the units and senior officers involved. Warren knew O'Rorke from when the former had been his brigade commander. Hazlett's battery had supported Warren's brigade at the Second Battle of Bull Run.

2. Create Shared Understanding. Warren was on Meade's staff as chief engineer and present for all major decisions. He had toured the lines with Meade and was, therefore, very familiar with Meade's concept for the defense. O'Rorke and Weed, having previously served with or under Warren, were also familiar with his tactical sense and were able to quickly respond to his instructions without requiring time consuming details. All officers involved quickly realized the danger posed by a Confederate attack on Little Round Top and responded energetically.

3. Provide a Clear Commander's Intent. Meade gave clear guidance but it was of such a general nature that Sickles' interpreted it for his own means. Other officers, including Sykes, Warren, O'Rorke and Weed, were forced to use their own initiative to recover from Sickles' interpretation, which threatened the whole left flank of the army. That these officers, all of whom, unlike Sickles, were professionally trained, were able to respond as Meade expected them to, demonstrates that Meade's intent was clear to them.

4. Exercise Disciplined Initiative. Meade's intent clearly provided latitude for initiative among his subordinates within certain parameters. These limits, while seeming obvious to the military professional, were

less apparent to political generals such as Sickles. While subordinates were granted freedom in deploying their units, this was limited in that no subordinate commander could place his unit in a position that may be optimal for the unit but less than optimal for the army as a whole, particularly without coordination with the effected units. Therefore, sickles displayed undisciplined initiative. However, Warren, Vincent, O'Rorke and Weed all demonstrated disciplined initiative, adjusting their decisions and actions within Meade's general concept of defense to a changing enemy situation that required a swift response. Vincent and O'Rorke displayed initiative at an even lower level, responding without orders from their immediate commanders to what they saw as an emergency situation. Their responses proved to be correct.

5. Use Mission Orders. Meade had only been in command of the Army of the Potomac for several days by the time of the battle. As such he typically issued general directives and appointed key subordinates to lead parts of the army in his absence. The use of such general directives led to situations such as Sickles' deployment, where Meade's intent was not followed. However, using the mission paradigm, the army commander responded by depending on key subordinates, such as chief engineer Warren, to translate his intent into action on the battlefield and respond to changes in the situation (such as an imminent enemy attack on an unoccupied Little Round Top).

6. Accept Prudent Risk. The uncertainty of the nature of the Confederate attack played a key role in this action. Sickles, unaware of the exact axis of the Confederate attack, chose to occupy a position more advantageous for his corps but less advantageous for the army as a whole. Then he compounded his imprudence by failing to coordinate his movements with the rest of the army. Meade quickly realized the uncertainty on his left flank and prudently sent reinforcements there from the Fifth Corps as soon as they became available. From these reinforcements, which were originally given relatively vague missions, subordinates on the spot were able to prudently change these missions into new assignments by taking the risk that the continued possession of Little Round Top by friendly forces was more important than any other tasks these units could have been doing at the time.

Section 2: Cases at Brigade/Regiment/Battalion Level

Nelson, Mission Command, and The Battle of Nile

John T. Kuehn, Ph.D.

The Battle of the Nile occurred off the coast of Egypt on the night of 1-2 August 1798. The Nile numbers among the most decisive naval engagements in modern history and remains a monument to the superior training, tactics, organization, and especially leadership of the Royal Navy of the period. This vignette looks at the supreme commander of the British Fleet at the Nile, Admiral Lord Viscount Horatio Nelson, Baron of the Nile. One finds all the elements of mission command employed by Nelson in his annihilation of the French fleet anchored not far from the Nile Delta at Aboukir Bay that night. Nelson left a profound legacy to mission command that still affects how navies, and especially the United States Navy, operate today. Nelson codified for the Royal and all other navies, the guiding principle of pre-battle centralized planning and decentralized violent execution in combat while at the same time underwriting the tactical initiative of his subordinates. His subordinates in turn relied in a similar fashion on the teamwork and initiative of their crews. Similarly, we find that Nelson's commander, Admiral John Jervis, also exercised a form of mission command in how he "controlled" his talented subordinate. The vignette thus demonstrates how mission command at all echelons of command can be a profound force multiplier.

* * *

The period prior to the actual battle has much to teach us about mission command in the Royal Navy, what Nelson himself called the "The Nelson Touch."

In 1798, France and Great Britain had already been at war with each other for five years after the French Republic beheaded King Louis XVI. Since then, the war between France and most of the rest of Europe had gone back and forth and the stalemate had only been broken in 1797 by the fabulous victories of a young French general in Italy named Napoleon Bonaparte. For much of 1797, Britain faced the French alone including a threat of invasion. The British disposed of this threat through battle and luck. The French had hoped to invade the British Isles proper using a combination of three fleets: the Spanish, Dutch, and French. However, British admirals had smashed the first two fleets at the battles of Cape St. Vincent (February 1797) and Camperdown (October 1797). As a final blow, a late season hurricane destroyed many of the French *bateaux* (light flat-bottomed boats) that were being built up for the invasion of England in late 1797, although the French fleet without the Dutch would have probably cancelled the invasion even without the storm damage. Napoleon proposed to the Directory (French National Government in 1797) that he instead attack Great Britain by the indirect route by seizing Malta and Egypt to threaten Britain's long and vulnerable line of communication with its most important colony in India. The Directory, wanting to get rid of the political threat posed by General Bonaparte, agreed and dispatched him with over 44,000 troops and the entire French Mediterranean Fleet to accomplish this task. Bonaparte captured Malta and then proceeded leisurely to Egypt where he landed unmolested and defeated the local Egyptian *Mameluke* Army at the Battle of the Pyramids (July 1798).

In 1797, Horatio Nelson was one of the youngest admirals in the Royal Navy and with Admiral John Jervis commanding the British Mediterranean Fleet; Nelson won everlasting glory at the Battle of Cape St. Vincent. It was here that the first elements of Nelson's decentralized command style are first clearly seen. Nelson, disobeying the famous Standing Battle Orders that were rigid tactical doctrine of the Royal Navy of that day, wore (maneuvered) out of the battle line without orders and sailed straight for the middle of the Spanish column where the most powerful ships sailed, including the massive flagship *Santissima Trinidad* with 140 guns. Nelson's aggressiveness and ability to act independently had been known in the fleet but now they were on display for all to see. Jervis, aboard the *Victory* with 100 guns, saw Nelson now engage seven enemy battleships with his one. He approved the action and signaled the *Collingwood* and *Excellent* to support him. Jervis then sent out the same signal to the remainder of his ships that Nelson himself sent from *Victory* at Trafalgar, "Engage the enemy more closely." Obviously, Jervis had created a shared understanding with subordinates. Nelson followed this success with his first defeat in an amphibious assault on the Spanish garrison at Tenerife in the Canary Islands, where he lost his right arm.

Nelson's record at Cape St. Vincent ensured his assignment to the critical theater of the war in 1798 again under Jervis. Nelson raised his flag on *Vanguard* in late March and set sail for the Gulf of Cadiz. Jervis's confidence in Nelson was unbounded. As soon as Nelson arrived to join the fleet on blockade duty of Cadiz, Jervis detached him on an independent command with a small squadron to enter the Mediterranean and keep an eye on the French fleet in Toulon. Jervis forwarded another eleven battleships to Nelson in May 1798, an unprecedented command for such a junior admiral which caused much grumbling amongst the many admirals senior to him without sea command. Jervis instructed Nelson to search for Bonaparte's invasion force, suspected to be bound for Egypt. Nelson's plan was simple, intercept the invasion fleet and destroy Napoleon and his army at sea.

Jervis's faith in Nelson was rewarded but not right away. Nelson's impatience almost did him in but he was tenacious in pursuit of his quarry. Nelson had been off Toulon prior to his reinforcement but had sailed off and then had his small squadron scattered by a storm. Napoleon had departed on 19 May, escorted by the French fleet under Admiral Francois-Paul Brueys, had departed. Nelson was desperately short of frigates (he had only three) to provide him intelligence and he guessed Napoleon's destination was Naples. Napoleon instead went to Malta and quickly conquered that small island. Nelson realized his mistake and determined that Napoleon's next objective was Alexandria, Egypt. He rushed to and arrived off Alexandria on 29 June and found nothing. Brueys and Napoleon had taken a different route via Crete and sailed far slower than Nelson imagined. Nelson missed a great chance by not waiting off Alexandria, instead second-guessing himself and sailing north to Turkey to seek the French fleet. While Nelson sailed north, the French arrived and began to debark their troops. It seemed that Nelson had lost the game of cat and mouse and missed a golden opportunity to destroy both a French Army and a French Fleet.

Nelson did not give up. Off Sicily, he learned of his mistake and doubled back to Alexandria. He arrived late on 1 August and found the transports empty but Brueys' 13 battleships and many smaller warships lay anchored close in to shore in the shallow and treacherous shoal water of Aboukir Bay. Brueys had unwisely sent half of his gun crews ashore to assist Napoleon with the land campaign. He also thought himself unassailable so close in to the shore. The final nail in his coffin was the late hour of the day. Surely Nelson would not attack in such dangerous waters in the dark. Nelson, demonstrating his intuitive grasp of the weakness of the French, instantly decided to attack.

Nelson had already planned for this moment and so could make the decision almost instantaneously to attack the vulnerable French ships. To understand how he prepared, one must go back many years. First, Nelson actively fought for the welfare of his sailors from the beginning of his career until his death in combat in 1805. This was known by the sailors who served under him and generated a confidence in his decision-making and leadership. Second, Nelson's rapid and seemingly "snap" decision-making in combat was not so much luck or intuition, although these certainly played a role, but were in the words of one historian, "achieved by decisions made in the quiet of his cabin." In other words, Nelson prepared himself intellectually at all times for decision making, tactical and otherwise, through reflection during the often lengthy downtime that one experiences at sea.

For this battle, Nelson had the complete confidence of his 14 battleship captains and this was rewarded by his confidence in them. Nelson also had the advantage of having some of the best captains, ships, and crews in the entire Royal Navy under his command at this battle. The 10 battleships that St. Vincent sent him under Sir Thomas Troubridge, who had served as a midshipman with Nelson, were later referred to by Nelson himself as "the finest Squadron that ever sailed the Ocean." He also knew most of their captains personally after having served with most them in combat, three he knew by reputation alone. He soon remedied this deficiency by proactive efforts. After barely missing the French at Alexandria and as he sailed about looking for them, Nelson instituted the policy of bringing his captains aboard the flagship individually to eat with him and share ideas. In the words of a contemporary observer, "he would fully develop them to his own ideas of the different and best modes of Attack…they could [as a result] ascertain with precision what were the ideas and intentions of their commander without the aid of further instructions."

Nelson brought his four most senior captains aboard his flagship HMS *Vanguard* on 22 June and asked them for their opinion on which way the French might have gone. They all voted for Alexandria as they were all thinking along the same lines as Nelson. This shows how interested the admiral was and how much he trusted in his subordinates' judgment. Just before turning the squadron back to Alexandria on 17 July, Nelson brought all 14 captains to the flagship to ensure commonality of purpose and a shared understanding for what he expected from them in the battle he was convinced would occur soon if he could only find the French Fleet. Nelson's only real "error" prior to the battle involved his reluctance to formally appoint a second-in-command since he was actually junior in terms of service time to several of his subordinate captains and did not want to upset the cohesion of the team. Perhaps he was thinking that if he was killed (a distinct possibility given his habit of being in the thick of battle), they would all continue to do their duty out of sense of obligation to him rather than have their own fighting focus diminished by petty jealousies.

As the British fleet closed on the French in Aboukir Bay, Nelson had 14 battleships to 13 of the French but the French had the heavier weight of gunnery in bigger ships and bigger guns. Upon sighting the enemy fleet, Nelson issued signals 53 and 54 from the Royal Navy's official signal books. The first simply alerted all the crews of his fleet to "prepare for battle." The second directed them to be ready to anchor at a moment's notice by the stern of the ship, a rare maneuver requiring the rigging of a very large anchor and cable at the rear of each ship. This order meant, as well, that each captain and his crew would have to anchor under enemy fire while unable to fire their guns and fully occupied with furling (taking down) all of their sails. Obviously both a cohesive team and shared understanding had to be present for this unprecedented tactic to be executed properly. Each captain, and more importantly their crews, knew exactly what this maneuver meant and welcomed it. Further, each captain received the order and executed it without signaling back to Nelson for any further guidance. This was only possible because they all knew their commander intimately, they knew their ships (their "weapons system") intimately, and they knew that their enemy was unprepared both for battle and especially for this unprecedented maneuver. Nelson intended for his anchoring to take place in a single battle line between the French line and the open sea.

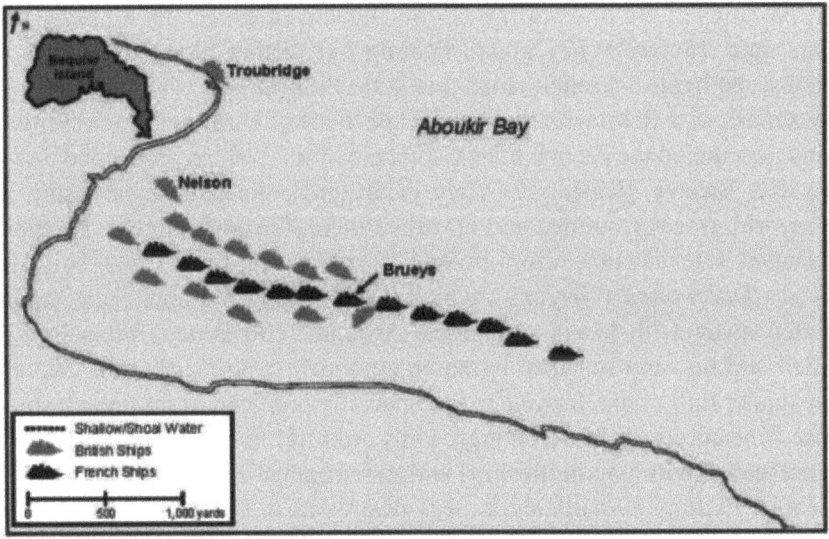

Battle of the Nile – British double envelopment of the French Fleet.

The second, we might say higher, level of mission command emerges in what the captains then did in response to Nelson's guidance to them about relying on their own initiative. Since he was in the rear aboard *Vanguard* with (74 guns), Nelson effectively gave his captains free reign when he sent the signal to his ships at 1730 "to form line as most convenient," leaving it to their discretion how to approach the French fleet. The execution was not without some flaws. Captain Troubridge on

Culloden (74 guns) ran aground just north of the French on the approach, but this accident served to let the ships behind know where the hazard lay. Several ships performed the anchoring maneuver badly and paid with heavy casualties as a result of being out of position but Nelson's other captains served him well. The aggressive captain of the lead ship *Goliath* (74 guns) was Captain Sir Thomas Foley. He instantly made the decision to go behind the first French ships and inside the French line. About half the other captains followed his initiative while Nelson signaled the remainder to follow the *Majestic* (74 guns) on the northern side (see figure). This resulted in a double envelopment of the French ships in the first part of the line. Their mates further down the line were anchored solid and could not help them.

To make matters worse, some of the French ships had not strung cable between their neighbors to counter the known British tactic of "breaking the line" and then shooting broadsides longitudinally into the aft and bows of the anchored French ships. Again, taking local tactical initiative, several British ships of the line performed this maneuver, especially *Leander* (50 guns) and *Alexander* (74 guns), dealing out further bloody devastation against the hapless French ships that were their targets. The night sky of Alexandria was lit up with the din and spectacle of burning French ships as Nelson pounded the French Fleet to pieces while his ships proceeded down the line. Brueys' flagship *L'Orient* (120 guns), the largest battleship in the world, was surrounded and absorbed incredible punishment. *L'Orient* managed a devastating return broadside against the first British ship she faced, *Bellerophon* (74 guns), which, mishandling the anchoring tactic, drifted away with heavy casualties. Around 2100 Brueys himself was killed and his ship set afire by some combustible grenades that Captain Alexander Ball of the *Alexander* had specially prepared to set French ships afire if he could get close enough. Only Admiral Pierre Villeneuve in the rear escaped with two battleships and two frigates. The rest of the French Fleet, including 11 battleships, was destroyed or taken. Nelson, in the thick of the fighting as usual, received a nasty head wound.

The battle continued through the night. The last two French ships, *Tonnant* (80 guns) and *Timoleon* (74 guns) surrendered around daybreak. Nelson had not lost a single ship and had less than 900 casualties, most of these in ships that had mishandled the anchoring maneuver. By destroying the French fleet, he had effectively checkmated France's strategy by stranding Napoleon's army in Egypt. When Napoleon attempted to fight his way through Ottoman Turkish territory in order to gain passage to Europe or even India, he was confounded again by that other great naval hero Sir Sydney Smith who successfully led the defense of Acre on his own initiative, forcing Napoleon to turn back to Egypt. Napoleon abandoned his army in 1799 and returned to France where he took over the government in a *coup d'etat*.

After the Nile, Nelson's fame reached unparalleled heights. The Sultan of Turkey presented him with an elaborate diamond decoration that Nelson wore on his large bicorn hat. The King rewarded him with the title of Lord Nelson, Baron of the Nile, and his income doubled when he received a ten thousand pound gift from the grateful East India Company. He was the darling of the press and had reached what today is known as "rock star" status. All of the complaining against both Nelson and Jervis ceased, Jervis feeling more than vindicated by his decision to give the young admiral a chance to strike a key blow. Many historians consider the Nile to be Nelson's most important victory, strategically and tactically. However, for Nelson, his captains and crews at the Nile remained forever in his heart and in his correspondence as his "band of brothers."

For Further Reading

Terry Coleman. *The Nelson Touch*. Oxford, UK: Oxford University Press, 2002.

Arthur Herman. *To Rule the Waves: How the British Navy Shaped the Modern World*. New York: Harper Collins, 2004.

Roger Knight. *The Pursuit of Victory: The Life and Achievement of Horatio* Nelson. New York: Basic Books, 2005.

Tom Pocock . *Horatio Nelson*. New York: Alfred A. Knopf, 1988.

John Sugden. *Nelson: The Sword of Albion*. New York: Henry Hold and Company, 2012.

The Six Principles of Mission Command
  1. Build Cohesive Teams through Mutual Trust
  2. Create Shared Understanding
  3. Provide a Clear Commander's Intent
  4. Exercise Disciplined Initiative
  5. Use Mission Orders
  6. Accept Prudent Risk

Mission Command in the Battle of the Nile case

1. Build Cohesive Teams through Mutual Trust. All units of this operation were veteran formations, most of which had worked together in the past, including Nelson's superior in command Admiral John Jervis, Lord St. Vincent.

2. Create Shared Understanding. Admiral Nelson shared his vision both individually and collectively with his captains. In turn, the captains of at least 11 ships seem to have shared this vision with their crews given their crews' flawless execution of the dangerous stern anchoring maneuver under fire.

3. Provide a Clear Commander's Intent. Nelson provided intent over time, such as at one on one dinners in his cabin with individual subordinates or larger meetings while at anchor with several captains at a time. His clear objective focused on the destruction or capture of the entire French Fleet while anchored or afloat. Nelson's signals to his captains at the outset of the battle set the tone and intent for the entire engagement.

4. Exercise Disciplined Initiative. Once Nelson sent his final signal to the entire fleet, he gave his captains free reign. One of them initiated the famous "doubling" tactic inside the French line in the most dangerous shoal waters of the bay, which Nelson supported with his own maneuvers and signals.

5. Use Mission Orders. British orders came from a pre-published list in the Royal Navy Signal book as informed by the Standing Battle Orders and, most importantly, the pre-battle briefings of the commander. These approximated today's mission orders, his presumption being that he need not provide detailed guidance since he expected his captains to all have mastered the pre-planned responses implied in signals and the Battle Orders. The commander had a wide range of signals to choose from to add additional clarity and flexibility. The constraints of naval combat using signal flags mandated clear, cogent orders that would be understood immediately by a subordinate captain.

6. Accept Prudent Risk. By the standards of any other navy, the risk in attacking at night, in shoal water, under fire, while performing one of the most difficult anchoring maneuvers would normally be characterized as very risky, even foolhardy. However, Nelson's confidence in the seamanship of his captains, and their confidence in their crews and in Nelson mitigated this risk. Further, in the Royal Navy, while the risk for this sort of thing would probably have been considered a bit beyond the norm, Nelson's methods at this battle were universally applauded by his fellow officers and by the senior leadership. After all, he had a slight superiority in numbers of ships over the French. It would have been considered a dereliction of duty and possibly a courts martial offense for a British officer to have NOT attacked the French, although possibly not as quickly and rapidly as Nelson did.

Assault on Queenston Heights, October 1812

Richard V. Barbuto, Ph.D.

In the dark early hours of a cold October day, a small group of American infantry conducted an opposed river crossing to capture a gun that threatened to defeat the major US operation – the invasion of Canada. Despite the ultimate failure of the invasion, the activities of the men of this intrepid band cleared the way for their comrades to cross the treacherous waters of the Niagara River in relative security.

On 18 June 1812, the United States declared war on the British Empire. President James Madison was determined to secure neutral shipping rights and to stop the forced impressment of American sailors on the high seas by Royal Navy captains looking to replenish their crews. Congress had taken steps in January to prepare America for war by authorizing the formation of new regiments of infantry, artillery, and light dragoons. However, selecting hundreds of new officers and recruiting thousands of new soldiers took time. Madison hoped to capture the key city of Montreal on the Saint Lawrence River in 1812 and to follow up by seizing the fortress city of Quebec the following year. However, execution fell short of intentions. The first invasion of Canada, across the Detroit River in July, ended up with the surrender of 2,400 regulars and militiamen in August.

The second invasion attempt, under New York militia MG Stephen van Rensselaer, was expected to secure the Niagara River and to stand ready for continued operations in support of the main invasion to take Montreal scheduled for November. The Niagara River is a 37-mile long strait emptying the waters of Lake Erie into Lake Ontario. The area is cut by a steep-sided 180-foot tall escarpment that hinders north-south movement. Because of the rapids leading to Niagara Falls and the steep gorge north of the falls, the river was crossable for only a portion of its length. The British had established artillery positions covering the likely crossing sites to prevent a US assault.

MG van Rensselaer, who had no military experience himself, depended heavily upon his distant cousin, LTC Solomon van Rensselaer. Solomon had served as a junior officer in "Mad Anthony" Wayne's successful war against the Indians of the Old Northwest in 1794. Solomon had been shot in the lungs but survived his near-fatal wounds. He left the regular army and joined the New York militia. Now back in the field, he planned the operation that would land an army of 4,000 militiamen and regulars on Canadian shores to secure the invasion route.

MG van Rensselaer chose to cross the Niagara River near the village of Queenston at the foot of the escarpment. At this point, the flat ground on the top of the escarpment was called Queenston Heights. The river spilled out of the Niagara Gorge at about four miles per hour. The flow, generally north, was characterized by back currents and swirling eddies that appeared and disappeared without warning. The river's banks were steep and between twenty and forty feet tall. While for most of its length the river came right up to the banks, at intervals there appeared narrow gravel beaches allowing the boats to land and the soldiers to disembark. The best landing place was at the foot of Queenston Village where a winding road connected a small dock to the river road atop the bank.

The commander of British forces along the Niagara River was none other than the brave and daring MG Isaac Brock who had forced the surrender of the garrison at Detroit. Despite being greatly outnumbered by American forces across the Niagara River, Brock had built a balanced defense and had established a system by which the local militia could quickly be summoned in response to an American invasion. Brock stationed two companies of regulars of the Forty-Ninth Regiment of Foot at Queenston with one company in the village and the other atop Queenston Heights. He placed a nine-pounder gun at the Queenston Landing and an eighteen-pounder gun in a redan (fortification) about two-thirds of the way up the escarpment. The redan was little more than a low stone wall on a narrow ledge on the slope of the steep hill. British and Canadian gunners in the redan had a clear field of fire across the river to the potential embarkation points and could fire at boats as they transited the dangerous waters. However, because the slope leading to the redan was so steep, the gun could not be depressed to sweep the approaches to it from the low ground below. The Americans were well aware of both guns and judged the artillery piece in the redan as the most serious threat to the river crossing.

The command team of the van Rensselaer cousins planned to cross the river with 13 boats. All 13 boats together could carry 300 soldiers in each wave. The army secured the services of civilian boatmen to pilot the boats. The boatmen would man the tillers while eight soldiers in each craft pulled the oars. Van Rensselaer estimated that it would take about 10 minutes to cross the river. Given up to 10 minutes to load and unload soldiers, the entire round trip would consume about thirty minutes. There were four thousand soldiers plus artillery and wagons of supplies and ammunition to cross. It was possible, even likely, that some boats would be lost while crossing, either sunk or swept downstream by the current. Neither the general nor his chief planner saw these parameters and constraints as serious. While there were additional boats available for use, the general did not order these brought to the embarkation site.

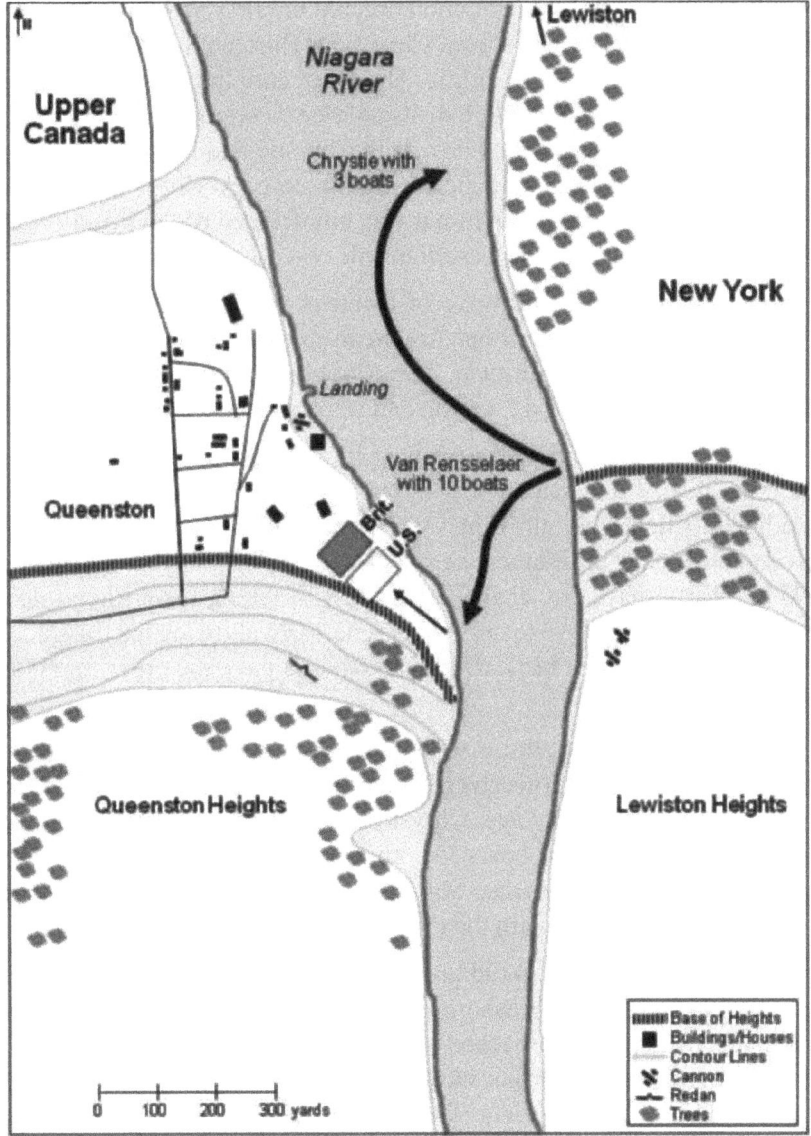

Figure 1. Initial US crossing and confrontation with British Infantry, 0400-0530, 13 October 1812.

The American army in its earliest wars demonstrated a profound tension and lack of trust between its regular and militia components. This issue nearly scuttled the crossing. General van Rensselaer designated

Solomon to lead the attack and to be its tactical commander in Canada. Regular Army LTC John Chrystie, who was senior in rank to Solomon van Rensselaer, had to agree to follow the militiaman's plan or not to participate in the operation. Solomon ordered the first wave of 300 men to be composed of 150 regulars from Chrystie's Thirteenth Infantry and 150 of the best-trained New York militia. The Thirteenth Infantry was a newly-formed unit and it is arguable that its soldiers were no better trained or led than the best militia companies. John Chrystie and his men joined the invasion only hours before the boats were to embark. Chrystie may have seen the crossing site in daylight but was uninformed of the details of the topography and roads on the Canadian side.

The mission of the first wave of invaders was to seize the eighteen-pounder gun in the redan. Other American artillery could not range this large gun because of its altitude. The American guns would engage the nine-pounder at Queenston Landing and strike targets of opportunity.

Chrystie started loading the boats at about 0330 hours on 13 October. While he clearly understood that he was to fill only half the boats, he nonetheless filled them all. When van Rensselaer arrived with his small staff and the picked militiamen, he was understandably angered. However, he knew that to have any chance of success of taking the artillery piece, the first wave would have to cross in darkness. Swallowing his frustration, van Rensselaer ordered his staff into a boat and gave the order for all the vessels to push off.

The crossing did not occur without mishap. LTC Chrystie's boat and two others were swept downriver. Perhaps the pilots lost their nerve and covertly sabotaged the voyage. Chrystie's party was making no headway and he ordered the three boats to return to the New York shore to try again. Solomon van Rensselaer was entirely unaware that his second-in-command was no longer with the crossing party.

The 10 remaining boats did not head for Queenston Landing. Instead their pilots brought them ashore on a shale beach at the base of the escarpment. As they approached shore, however, British sentries fired down from atop the riverbank into the darkness, killing LT John Valleau and wounding several soldiers. The sentries fled, no doubt reporting the American landing. The infantrymen, now about 225 in number, slowly climbed the steep banks in the darkness. CPT John E. Wool, not finding Chrystie or van Rensselaer, took command and formed the men in a column facing the redan, preparing the men to scale the heights. Arriving in the last boat, van Rensselaer was slow to join the assault party. In the darkness, van Rensselaer sent his judge advocate, Stephen Lush, to the head of the column with orders to begin the ascent. Both Wool and van Rensselaer wondered at the whereabouts of John Chrystie.

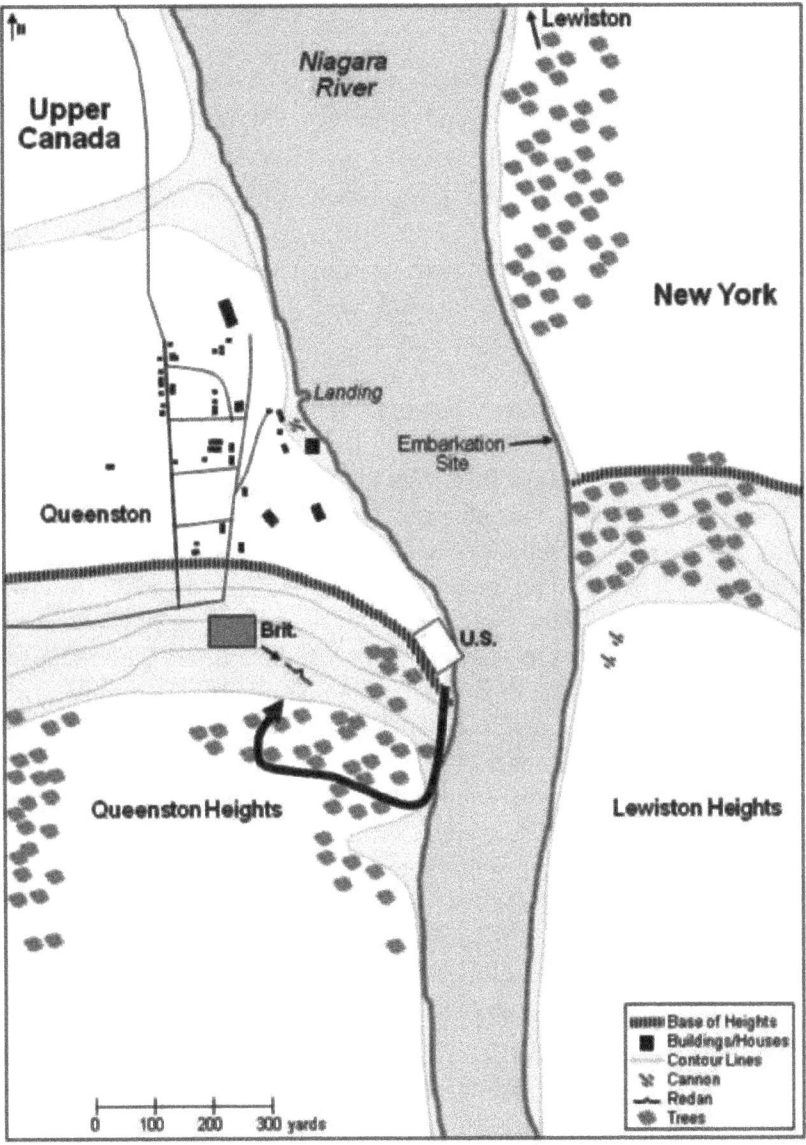

Figure 2. CPT John Wool's detachment ascends Queenston Heights, 0530-0700, 13 October 1812.

Meanwhile, the local British Commander, CPT James Dennis, roused his troops from their slumber. The men had been resting in their uniforms with weapons at their sides and were soon ready to march. The two British guns, spying Chrystie's boats in the river, opened fire. Not knowing where the Americans might land, Dennis kept most of his men near the village. He led a company of grenadiers and some militiamen at hand toward the reported landing at the base of the escarpment. In the darkness, the two bodies of infantrymen, British and American, stumbled into one another and opened a furious firefight.

The Americans eventually drove off the grenadiers but not before losing a number of men. Six of 11 officers were down, two mortally wounded. Two of the five company commanders were out of the fight. Wool had turned sideways to give an order to his company when a British musketball passed through both buttocks. He fainted momentarily but was soon revived. Van Rensselaer was even less lucky. Leading from the front, he received five bullet wounds to his legs and a heel. Knocked to the ground, his white trousers red with his blood, the weakened colonel ordered Lush to wrap him in his greatcoat so that the soldiers would not view him and lose their courage. Van Rensselaer sent Lush up and down the lines searching unsuccessfully for Chrystie. Van Rensselaer understood that he could no longer command and he ordered the survivors to gather at the edge of the riverbank to await Chrystie or reinforcements. He did not know that in the emerging light, British artillery had discovered the embarkation site. The eighteen- and nine-pounder guns were slowly pouring solid shot and spherical case shot against the New York shore. The Americans were unaware of this new munition, later named shrapnel after its British inventor. They mistakenly thought that grapeshot was blowing through the ranks of soldiers trying to enter the boats. The returning boats brought wounded and dead with them, the sight of the casualties having the expected effect on raw troops. Some of the civilian pilots abandoned their boats and their duties. The American crossing plan was fast falling apart. A crisis had arrived and 200 American soldiers were alone on a foreign shore with no boats to return them.

Wool joined van Rensselaer to assess the situation. It was clear to both leaders that their party would be taken prisoner unless the redan gun was silenced. Wool was determined to complete the mission, to lead his shaken soldiers once again into danger with no guarantee and little likelihood of success. Wool, of course, had scant appreciation for the topography. Van Rensselaer directed militia LT John Gansevoort to guide the column, not up the face of the escarpment, but into the Niagara Gorge. There they would be concealed from sight as they picked their way skyward. Van Rensselaer also directed Lush to bring up the rear of the column with orders to shoot anyone who shirked his duty. As the sun's rays broke the darkness, the intrepid band began their improbable task.

Wool's detachment climbed single file up a narrow trail in the face of the gorge, slowly ascending more than 200 feet. While they progressed, American artillery found and silenced the nine-pounder at Queenston Landing. Four more boats put into the water, contending only with the redan gun. However, these boats did not row toward van Rensselaer's landing site, but instead headed down river away from the redan battery in an attempt to land north of the village of Queenston. The four boats closed onto the Canadian shore as British infantry 40 feet above them, fired directly into the craft. LTC James Fenwick had no choice but to surrender as his men were cut to pieces. However, MAJ James Mullany gathered several men about him. They placed a few wounded soldiers into one of the boats and managed to push off, eventually regaining the New York shore.

MG Brock had heard the cannon fire to the south and departed Fort George riding quickly to Queenston. Once there he conferred with CPT Dennis to assess the situation. Satisfied that this was the main American attack, Brock issued orders for more regulars and militia to concentrate at Queenston. He visited the redan battery to encourage the crew and to get a better view of the American shore. He then descended the escarpment to direct the actions of the newly arriving troops.

While Brock was in Queenston, Wool and his party arrived at the summit of their climb. Wool led his men to the brow of the escarpment and found the redan battery below. Despite his painful wounds, CPT Wool led the line of troops downward. Seeing the oncoming enemy, the artillerists abandoned their gun and withdrew down the slope. Finally, the menacing cannon was out of action. Wool disposed his men to defend the redan from all directions. He was not about to lose his prize.

Sometime during the early hours of battle, CPT Dennis had ordered the company of British regulars that were stationed on Queenston Heights to reinforce the troops in the village below. Had he not done so, it is probable that they would have been available to push the Americans off the Heights and secure the eighteen-pounder, keeping it in the fight.

View from the New York shore toward LTC Van Rensselaer's landing site. The Niagara Escarpment and the Brock Memorial are in the background.

Author's collection.

Perhaps more impetuous than prudent, Isaac Brock now made a fateful decision. He was determined to recapture his lost gun which he believed to be key to the defense. Brock galloped through Queenston gathering regulars and militiamen about him. He led them south through the village to the base of the steep slope of the escarpment. To the Americans in the redan, there was no mistaking the general resplendent in his gold-trimmed scarlet coat. Brock dismounted and sword in hand advanced at the front of a thin line of British and Canadian soldiers. A shot hit him in the hand, yet he continued his climb, urging his men forward. An unknown American soldier moved forward from the redan, took careful aim, and put a bullet in Brock's chest inches from his heart. Brock collapsed and his attack with him. In a minute the brilliant conqueror of Detroit lay slain.

View from the redan position, looking toward the New York shore.
Author's collection.

Over the next several hours, the Americans enjoyed a brief respite and as many as 1,000 crossed the river now that the two guns had been eliminated. Wool was evacuated as was Solomon van Rensselaer. Chrystie and Mullany managed to cross the river and take their commands atop Queenston Heights. LTC Winfield Scott assumed overall tactical command on the heights. Ultimately, though, the Americans forfeited what their intrepid leaders, John Wool and Solomon van Rensselaer among them, had gained. Eventually the militiamen still in New York saw a column of British marching from Fort George to the sound of the guns. They heard the fierce yells of native warriors allied to the British who were sniping at their comrades in Canada. They saw dozens of dead and wounded pulled out of returning boats. Despite the pleas of MG van Rensselaer and their militia officers, upwards of 3,000 militiamen invoked their constitutional right not to be ordered out of the country against their will.

Brock's second in command, MG Roger Hale Sheaffe, advanced by a far road to the top of the heights. He led a determined attack that overpowered Scott's line. Several Americans were seen plunging down the sides of the gorge to their deaths. Others attempted to swim the Niagara. None succeeded. Scott surrendered more than nine hundred men. Sheaffe eventually released the militiamen, but the regulars were marched into captivity.

MG van Rensselaer resigned his commission. He returned to his estates in Albany and to a hero's welcome. Many understood that this defeat was due more to the general unpreparedness of the army than to his inadequacies. In 1813, van Rensselaer ran for governor of New York, losing by fewer than 4,000 votes. After the war he started a college that is now known as Rensselaer Polytechnic Institute. Solomon van Rensselaer unsuccessfully vied for a command in the regular army despite never fully recovering from his wounds. John Ellis Wool, whose unwavering pursuit of the mission set up the conditions for victory, continued in the Army after the war. He rose to the rank of brevet major general for his gallant leadership in the war with Mexico. Wool served actively in the Civil War until his resignation in 1863 with the rank of major general.

For Further Reading

Robert Malcomson. *A Very Brilliant Affair: the Battle of Queenston Heights, 1812*. Annapolis: Naval Institute Press, 2003.

Pierre Berton. *The Invasion of Canada, 1812-1813*. Boston: Little, Brown and Company, 1980.

George F.G. Stanley. *The War of 1812: Land Operations*. Canadian War Museum Historical Publication No. 18. Macmillan of Canada, 1983.

Benson J. Lossing. *The Pictorial Field-Book of the War of 1812*. Facsimile Edition. Somersworth: New Hampshire Publishing Company, 1976.

The Six Principles of Mission Command

1. Build Cohesive Teams through Mutual Trust

2. Create Shared Understanding

3. Provide a Clear Commander's Intent

4. Exercise Disciplined Initiative

5. Use Mission Orders

6. Accept Prudent Risk

Mission Command in the Queenston case

1. Build Cohesive Teams through Mutual Trust. The raw recruits of the Thirteenth US Infantry followed their officers in a dangerous, nighttime opposed river crossing. Despite the loss of many of their officers, none shied away from a perilous climb up the steep sides of the Niagara Gorge. They assaulted the redan and defended it against a determined attack led by the most well-known and courageous of enemy leaders. These actions were certainly the result of mutual trust between officers and men.

2. Create Shared Understanding. Solomon van Rensselaer, John Wool, and the other officers understood that putting the redan gun out of operation was critical to the success of the mission. If they failed, the invasion would very likely fail as well.

3. Provide a Clear Commander's Intent. While the historical record is not entirely established, it appears that General van Rensselaer, through LTC van Rensselaer, provided every officer in the Thirteenth Infantry an understanding of the critical nature of their assignment.

4. Exercise Disciplined Initiative. Even when Wool and Solomon van Rensselaer were separated in the darkness and Chrystie was nowhere to be found, the officers and sergeants formed up the troops to prepare for an assault. They quickly responded to an attack by veteran British troops. Wool, with the assent of the surviving officers, and despite an incomplete appreciation of the situation, decided to continue the dangerous mission.

5. Use Mission Orders. Van Rensselaer and Wool changed the plan on the spot by deciding to enter the gorge rather than to attack the redan directly. Wool and his men remained focused on the goal, to capture the gun, rather than adhere to the original plan.

6. Accept Prudent Risk? Wool did not know what lay in wait at the top of Queenston Heights. He did not know, although he must have suspected, that he was outnumbered and that reinforcements were not forthcoming.

However, he also understood that he might very well achieve surprise if he gained Queenston Heights and assaulted downhill to overrun the redan gun. Van Rensselear helped mitigate risk by directing Wool's forces into the gorge where they would be hidden from British view and from which they could ascend the heights on a concealed avenue of approach.

A Motorized Infantry Regiment Crosses the Meuse River, May 1940

John J. McGrath

As part of the German main effort in the French campaign, the 1st Panzer Division was one of the spearhead elements of a large armored force which advanced through the Ardennes Forest and reached the obstacle of the Meuse River less than three days after the start of the offensive. Instead of reorganizing along the river and waiting for less mobile infantry elements to arrive to force crossings, the 1st Panzer Division used its internal resources of boats and infantry to immediately cross the river and create a bridgehead. Then the regiment attacked and secured key terrain that made the French defenses untenable and allowed the German divisions to the north and south to cross the river after initial failures. Then the German forces advanced deep into the allied rear area and cut the northern third of the allied armies off from the rest by reaching the English Channel on 19 May.

In May 1940, the German and Allied forces had been facing each other for over seven months in a period known as the Phony War. While the French and British mobilized and prepared for a German offensive, Belgium and the Netherlands, although positioned in the direct path of any probable German offensive, remained neutral. Meanwhile, the Nazi state proceeded to finish the conquest of Poland, aided greatly by Soviet intervention in the later stages. In April 1940, the Germans conducted a risky campaign with only a small number of troops to occupy Denmark and Norway, as Hitler feared the British were about to occupy the latter country, which would cut German access to key mineral resources in Sweden. In the ensuing operations, Denmark and the southern part of Norway were quickly conquered but the Germans were still fighting and losing to an Allied expeditionary force at Narvik in northern Norway when the campaign in the west started. The German dictator, Adolf Hitler, had not intended to delay the German offensive in the west until May 1940. However, circumstances continually delayed the start of the attack. During this period, the German plan was continually revised. Planning focused on how best to use the limited available armored forces.

In the interwar years, the Germans had developed two key concepts that played a big role in the success of the 1940 campaign. The first of these was the creation of the panzer division, a combined arms force that emphasized the massing of tanks supported by other arms in a flexible organization capable of offensive action based on the initiative of commanders at all levels. The second development was the creation of a tactical air force, the Luftwaffe, whose operations were designed to provide responsive and aggressive close air support for the ground troops. The primary aerial weapons platform was the Stuka dive bomber, which, essentially, provided German mobile units with the equivalent of highly effective long-range artillery. These two elements were tested in Poland in September 1939. While the Germans enjoyed success in that campaign, there were many teething problems which the Wehrmacht (German Armed Forces) used the Phony War period to correct.

Ironically, as the 1940 campaign opened, the British and French had both more and better tanks than the Germans. However, their tanks were organized primarily in pure tank units which were employed primarily in an infantry support role. The few armored division type organizations the French and British fielded had been hastily organized after seeing the success of armored forces in Poland. In contrast, the Germans, although fielding generally inferior tanks, massed them in panzer divisions, supported those tanks with infantry and other arms, and equipped each individual tank with a radio set.

Despite the fielding of the panzer divisions, the armored forces remained only a small part of the overall German military organization, which was composed overwhelmingly of leg infantry divisions whose artillery and supply wagons were pulled by horses. The German Wehrmacht fielded only 10 panzer divisions in 1940 and eight motorized infantry divisions, in which horses were replaced by trucks. Accordingly, German planning for the campaign was primarily a matter of determining how best to use the limited number of armored and motorized divisions.

The plan ultimately adopted, massed seven panzer divisions and six motorized infantry divisions in the center of the front opposite the Ardennes forest in Belgium. While the offensive opened with a secondary assault on Holland and Belgium spearheaded by the remaining three panzer divisions and paratroopers, the main effort would advance quickly through the Ardennes, overwhelm the main French defensive line along the Meuse River, and then advance rapidly to the English Channel, cutting the Allied front in half and isolating any forces in northern France and Belgium. The plan's effectiveness was amplified because the Allies expected the Germans to repeat the 1914 Schlieffen Plan in which the German Army massed forces on their right flank. The British and French intended to send their best forces into Belgium to meet the expected German maneuver head on and fight it to a standstill. With the German main effort actually advancing further south in the Ardennes, the Allied plan had the effect of sending the best troops into the lion's mouth.

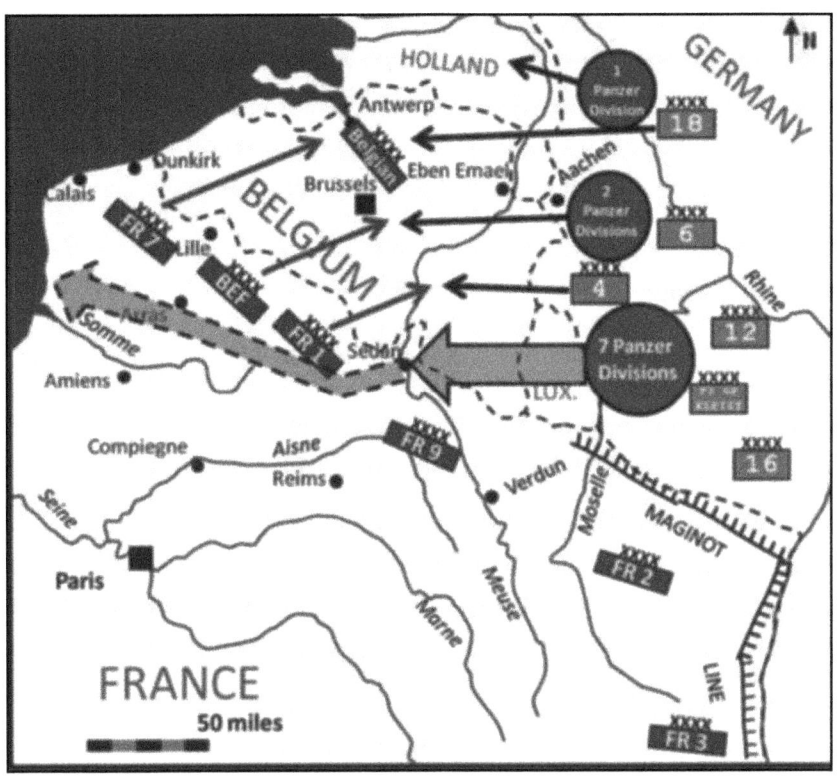

Figure 1. The Opposing Plans.

In the German Army of 1940 "mission command," known as *Auftragstaktik,* was more than a doctrine. It was the primary method of command throughout the force and had been since World War I with antecedents going back to the early 19th century. In the German model, mission command consisted of the issuing of short operational orders which gave subordinates a concise and clear idea of the commander's notion of the mission and a list of coordinating measures, such as unit boundaries. The order also provided all the support assets available to the higher commander to facilitate the subordinate's accomplishment of the mission. How the subordinate accomplished the mission was up to his own devices. To execute such an operational culture, the Wehrmacht depended on a corps of professional officers and NCOs which was the legacy of the small Army of the Interwar Period (1919-1939), and, at higher levels, a group of elite planners and operations officers - the General Staff officer corps. German leaders at all levels in 1940 were expected to provide immediate solutions to combat problems without waiting for guidance from higher levels but in accordance with the higher headquarters general concept for which mission the unit was to accomplish.

In 1940, panzer division organization was more tank heavy than it would be later in the war, but it was nevertheless a combined arms unit. After the Polish campaign, the number of panzer divisions was doubled, creating variations in organizational structure, particularly among the newer units, in the 1940 campaign. For illustrative purposes, the organization of the 1st Panzer Division, the subject of this study, is presented. In this division, the tanks were organized into a brigade of two regiments, each with two tank battalions, giving the division a total of four tank battalions. There was also a motorized infantry brigade that contained a motorized infantry regiment with three battalions and a separate motorcycle infantry battalion. The ratio of infantry battalions to tank battalions was therefore equal. The division also contained a motorized artillery regiment, a reconnaissance battalion, and an engineer battalion. For combat operations, the panzer division divided its elements into combined arms battle groups (originally called *Gefechtsgruppen*, later *Kampfgruppen*), typically joining battalions of armor, infantry and artillery together, along with smaller units of engineers and other support elements, under a single commander.

The Ardennes forest, into which these divisions would attack, could prove to be a major obstacle to the German advance, if defended by the Allies in strength. The terrain was rugged and cut by many small rivers. Armored and other vehicles were restricted in their movements to roads that could easily be blocked. The narrowness caused traffic jams. These terrain-based difficulties made allied planners discount a major German armored movement in the region. The Germans understood this and reinforced the idea by planning to start their offensive with a glider assault on the Belgian fortress of Eben Emael north of the Ardennes, misleading the Allied commanders as to where the German main effort was.

In May 1940, the 1st Panzer Division, commanded by LTG Friedrich Kirchner, was part of GEN Heinz Guderian's XIX Motorized Corps (Activated as the XIX Armee Korps and sometimes called the XIX Panzer Korps), a command that contained three panzer divisions and a separate motorized infantry regiment. Guderian was one of the German armored pioneers and his force was the spearhead of the main German effort in the campaign. All three of his panzer divisions contained four tank battalions. Starting on 10 May 1940 the three panzer divisions travelled along parallel routes through the Ardennes aimed at reaching and crossing the Meuse River, the expected main French defensive line, near Sedan as soon as possible. The 1st Panzer Division was in the center of the corps front, giving it the key role at the Meuse River.

Opposing the panzers initially was a covering force consisting of Belgian Ardennes Division a force of light and motorcycle infantry and the French 5th Light Cavalry Division (*5e Division Légère de Cavalerie* or 5e DLC). The latter was a new division consisting of a horse cavalry brigade and a mechanized cavalry brigade. It was hindered in its mission as it was unable to enter Belgian territory until the Germans invaded. As the campaign began and the German tanks crossed into Belgium, the French and Belgians fought tenaciously but were no match for the massed panzer force. Defending the main Meuse line was the 55th Infantry Division. This unit was a second tier French reserve organization which had been mobilized in September and manned with older reservists. Along the Meuse, the French had erected a fortified line of pillboxes and entrenched artillery positions. This line was a northern extension of the Maginot Line but was far less formidable. The French felt that the natural obstacles of the Ardennes and the Meuse would protect the defense in this sector and delay any German advance in the area.

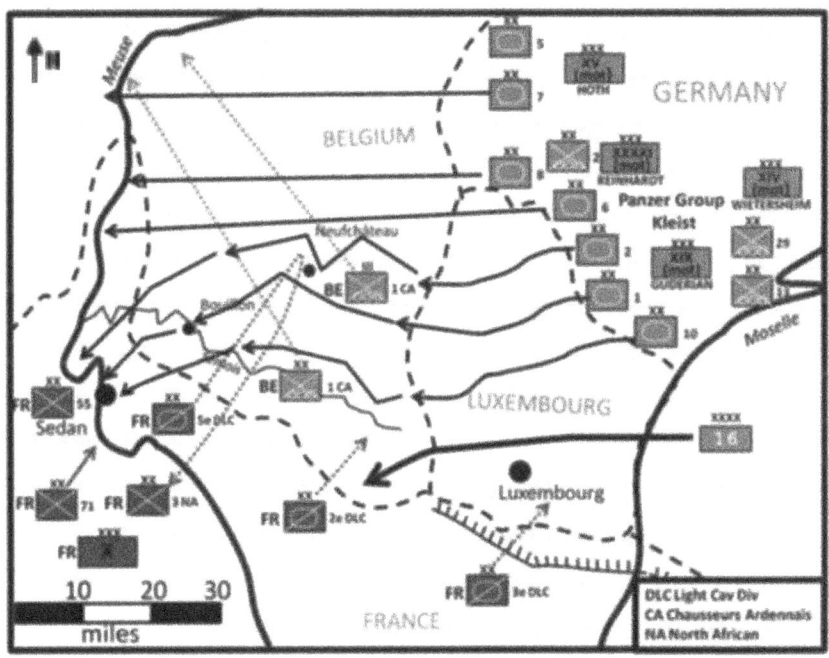

Figure 2. The German Advance Through the Ardennes, 10-13 May 1940.

In advancing through the 70-mile expanse of the Ardennes, Guderian expected his troops to reach the Meuse in four days and cross it on the fifth. His divisions were through the Belgian border defenses on the first day and reached the Semois River, the last major obstacle before the Meuse, on the morning of the third day. After one of the 1st Panzer Division's motorized infantry regiments the 1st Rifle Regiment attacked and cleared the town of Bouillon on the Semois that same day, advance elements of the 1st Panzer Division reached the Meuse that night, a day ahead of schedule. There the divisions reorganized for the crossing, with the motorized infantry to lead the way.

German Troops Advance on Sedan, 12 May 1940.
National Archives.

The German plan depended upon surprise. The massing of the panzer forces in the Ardennes sector could hardly have been expected by the Allied planners, so it was important that the maneuver only be revealed after the main French defensive line, which was along the Meuse River, had been ruptured. Otherwise, the allies may have been able to respond by sending reinforcements to the sector which could result in a failure to breakthrough. For this reason the German leadership down to the lowest NCOs in the 1st Panzer Division, were motivated to use initiative and daring to keep the advance moving forward.

The swiftness of the advance forced Guderian to adopt a river crossing plan used in a wargame the previous month as the order for crossing the Meuse. German staff officers merely changed time schedules and objectives to match with the current situation and sent the revised operations order to the units. In turn, the units sent similar orders to their subordinates. The simplicity of this method clearly showed the familiarity of the leaders in the corps and the 1st Panzer Division. Guderian commented in his memoirs on how well he knew the commanders of all his subordinate units. LTG Friedrich Kirchner, the commander of the 1st Panzer Division, had commanded the division's infantry brigade in Poland and had been the division commander since November 1939. The division chief of staff, MAJ Walther Wenck, had held his position in the Polish campaign. The 1st Rifle Regiment commander, LTC Hermann Balck, had held his post since November 1939. MAJ von Jagow, the 2d Battalion commander, had been in that position since December.

German infantry carry a raft in preparation for crossing the Meuse, May 1940.
National Archives.

The resulting corps operations orders were relatively simple. The warning order was issued at 1750 hours on 12 May and only consisted of five paragraphs and less than 200 words. The 1st Panzer Division, in turn, issued a terse warning order at 1845 hours. The warning orders adopted the exercise plan used the month before. The units would cross the river with units given the same tasks they had had in the exercise. The corps operations order was issued at 0815 hours on 13 May and was two and a half pages long. The 1st Panzer Division issued its order at noon. This order was five pages long and included an artillery fire plan and timetable. The mission statement was as follows: "1st Panzer Division… will be ready to attack at 1600 hours. After mopping up inside the Meuse bend it will push forward to the Bellevue-Torcy road. The division will proceed to attack the Bois de la Marfée heights and will push on to a line Chéhéry-Chaumont."

Panzer II and Panzer I in the west.
National Archives.

Engineer units moved forward and placed crossing equipment near the bank of the river. Arriving by trucks, the 2d Battalion, 1st Rifle Regiment, 1st Panzer Division, MAJ von Jagow commanding, reached an assembly area just short of the west bank of the Meuse at 0600. The battalion was designated to conduct the initial assault river crossing. Since the Germans expected tough French resistance, an extensive aerial and artillery preparation was planned before the river crossing. Two hours after the infantry arrived, this preparation began. The infantrymen rested behind the river bank for eight hours while most of the Luftwaffe's available attack bombers and all the artillery available in the corps bombarded the French defensive positions across the river.

At 1600 the 120 men of the battalion's 7th Company, commanded by 1LT Georg Feig, climbed over the sea wall overlooking the west bank of the river carrying assault boats that they then loaded into and rowed across the river. Feig had served as a platoon leader in the same company in the Polish campaign and was an original member of the regiment. The company was stopped at the far bank by a wire obstacle but quickly cut an opening in it using demolitions and advanced forward, bypassing forward bunkers and attacking them from the rear. The defenders, stunned by the aerial bombing and the swiftness of the German crossing, quickly surrendered. When French artillery fired at Feig's men from wooded high ground to the west, he moved quickly to silence the guns. The artillerymen abandoned their guns and fled before the advancing German infantry.

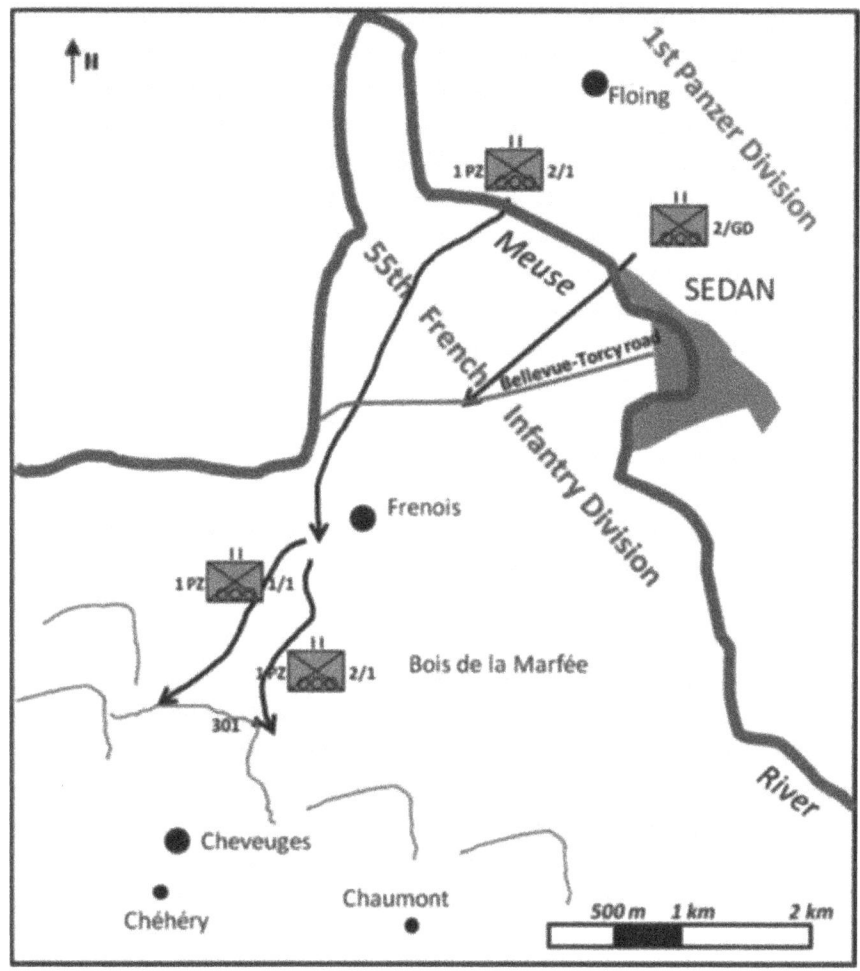

Figure 3. Crossing the Meuse at Sedan.

Followed by the rest of the battalion, the other two battalions of the regiment and the divisional motorcycle infantry battalion, Feig's men continued to advance and captured a key railroad crossing a kilometer from the river by 1730. By midnight, they had secured the high ground overlooking the crossing site three kilometers from the river, silencing the French artillery that had been stationed there. During the advance, Feig's men were joined at times by their battalion commander, MAJ von Jagow, and the regimental commander, LTC Balck, who encouraged the men and directed the arrival of reinforcements.

After the capture of the artillery, Balck assembled his battalion and company commanders and issued new orders, "We've got a small lane through the enemy. Let's break through." Balck later commented, "Something that is easy today can cost us rivers of blood tomorrow." The regiment advanced an additional kilometer, capturing all its objectives for the initial bridgehead including the key heights of Hill 301. The three battalions of the rifle regiment and the division motorcycle infantry battalion were now across the river. The Grossdeutschland (GD) motorized infantry regiment, attached to the division, crossed south of Balck's men. Once the engineers built bridges over the now secure crossing site, the tanks and other vehicles crossed the next morning. Additional divisions crossed the Meuse after the 1st Panzer. Guderian's corps and the other corps of the German main thrust then advanced westward deep into the Allied rear area, reaching the English Channel coast on 20 May and effectively winning the campaign. The British evacuated their troops from the trap at Dunkirk and France surrendered in June.

Feig remained with the 1st Panzer Division for most of the war rising to the rank of lieutenant colonel. He was awarded the Iron Cross Second and First Class for his exploits in France in 1940 and won the Knight's Cross as a company commander in Russia in December 1941. Jagow was killed two weeks after the crossing of the Meuse. Balck rose to be an army and army group commander by 1944. Guderian was relieved by Hitler in 1941 in Russia but returned a year later as the Inspector of Armored Troops. In 1944 he became the Chief of the General Staff, a position from which he retired several weeks before the end of the war.

The crossing of the Meuse shows mission command at its best. The Germans were able to adjust and synchronize their operations because of the flexibility afforded them by their system of issuing orders. This was particularly reflected in the adoption of a plan previously used in an exercise as the framework under which the river crossing was executed. In his memoirs, Field Marshal Erich von Manstein, the man who as Army Group A chief of staff had created the campaign plan used at Sedan, remarked on the German Army's use of mission command as follows, "Individual leadership was fostered on a scale unrivaled in any other army right down to the most junior NCO or infantryman and in this lay the secret of our success."

For Further Reading

Karl-Heinz Frieser with John Greenwood. *The Blitzkrieg Legend: The 1940 Campaign in the West*. Annapolis, MD: Naval Institute Press, 2005.

Heinz Guderian. *Panzer Leader.* Translated by Constantine Fitzgibbon. Washington, DC: Zenger, 1952.

Franz Kurowski. *Panzergrenadier Aces: German Mechanized Infantrymen in World War II.* Mechanicsburg, PA: Stackpole, 2010.

Erich von Manstein. *Lost Victories.* New York: Presidio, 1982.

The Six Principles of Mission Command

1. Build Cohesive Teams through Mutual Trust
2. Create Shared Understanding
3. Provide a Clear Commander's Intent
4. Exercise Disciplined Initiative
5. Use Mission Orders
6. Accept Prudent Risk

Principles of Mission Command in the Crossing of the Meuse case

1. Build Cohesive Teams through Mutual Trust. The division was cohesive. It had played a key role in the September 1939 Polish campaign. Its commander, LTG Friedrich Kirchner, had commanded the division's infantry brigade in that operation and had been the division commander since November 1939. The division chief of staff, Major Walther Wenck, had held his position in the Polish campaign. The 1st Rifle Regiment commander, LTC Hermann Balck, had held his post since November 1939. Jagow, the battalion commander, had been in command since December. Fieg, the company commander, had served as a platoon leader in the same company in the Polish campaign.

2. Create Shared Understanding. The Germans adopted an operations order for the river crossing which had been previously used in a wargame a month earlier. Accordingly all the leaders and soldiers had practiced a similar operation. And they understood clearly what the overall goal of the operation was: get across the river and penetrate as deeply as possible behind French lines.

3. Provide a Clear Commander's Intent. XIX Motorized Corps commander General Heinz Guderian designated the 1st Panzer Division as the corps main effort. The corps itself was the main effort of the entire German offensive. As such all leaders at all levels realized the importance of getting across the Meuse River as quickly as possible. The 2d Battalion, 1st Rifle Regiment, led the river crossing and its 7 Company was the spearhead.

4. Exercise Disciplined Initiative. Guderian was displaying initiative that bordered on the undisciplined. He pushed his troops across the river as soon as they got there because he feared higher headquarters would demand a pause to await the arrival of slow marching infantry that would reduce the surprise-created-by-speed factor that he felt was key to the overall success of the operation. At the lower level the German leaders

pushed across the river vigorously and continued to advance until the high ground overlooking the river over three kilometers away was seized. 7th Company was perhaps the most aggressive in these actions but it was not the only unit in the regiment to display this type of initiative.

5. Use Mission Orders. The German command system was based on mission orders. It was not just doctrine. It had become a philosophy and practice closer to a standard operating procedure. In their education and their exercises German Army officers and NCOs were inculcated with the principles of initiative, prudent risk, and mission orders.

6. Accept Prudent Risk. The Germans needed to get across the river before the French reinforced their defenses. There was the natural risk of any river crossing against prepared defensives (which there were) but the Germans mitigated these by using the Luftwaffe as mobile artillery. The 1st Rifle Regiment's crossing offers a number of examples of leaders accepting reasonable risk. Perhaps the best was LTC Balck's decision to continue to attack after taking out the French artillery position. He recognized the risk in penetrating deeper into French territory but saw the risk as greater if they put off the advance until later.

Corregidor

Triumph in the Philippines

Kendall D. Gott

In 1945, the strategically important task of clearing the Philippines main island of Luzon and capturing the port of Manila required the seizing of the island of Corregidor which guarded Manila Bay. In 1942, Japanese amphibious forces had taken very heavy losses in it capture and an airborne operation was conceived to prevent such casualties to the Americans. In a textbook display of good planning and excellent coordination between the services, the 503d Parachute Regimental Combat Team (PRCT) overcame difficult terrain and a desperate Japanese defense. As with all airborne operations, decentralized command, flexibility, personal initiative, and innovation were key elements in the successful conclusion of this mission.

By January 1945, the operations to recapture the Philippines from the occupying Japanese were in full swing with the northern island of Luzon as the main effort. The bulk of the Japanese were stationed there and the port of Manila would serve as a vital base to support future operations. The island of Corregidor guarded the entrance to the bay and any enemy forces left there could harass shipping and serve as a rallying point for any Japanese evacuating the mainland. The Americans had lost Corregidor after a dogged defense in 1942 and its recapture would be a great symbolic success.

GEN Douglas MacArthur outlined a general plan to the Sixth Army Commander, GEN Krueger, envisioning an airborne and amphibious assault of Corregidor following an intensive aerial bombardment and supported by naval gunfire. On 3 February, the Army G3 Operations tapped the 503d PRCT for the airborne drop and the reinforced 3d Battalion, 34th Infantry (3-34 IN) from the 24th Infantry Division for the amphibious landing. Both of these units were hardened veterans of the Pacific war and they would be ready for the proposed date of attack of 16 February. The 317th Troop Carrier Group of C-47s as well as naval forces quickly assembled their forces and needed logistical support within those short 13 days.

The decision to use airborne troops on Corregidor deserves examination. The island is but three and a half miles long and one and a half miles wide at its widest point. From above it looks like a tadpole with a large hill called Topside to the west forming the head and a long thin range of hills and ravines forming the tail to the east, called Bottomside. The area around Bottomside featured sandy beaches ideal for an amphibious landing and the airfield was located here too. However, the Japanese had used this area in their assault in 1942 and suffered tremendous losses as they had to advance up the hills and across the island to Topside. The Americans wanted to avoid that scenario by seizing the high ground first and then bring in reinforcements across the water. As with all airborne operations, surprise is vital, and the planners thought the defending Japanese were not expecting or prepared for an airborne assault. The Sixth Army staff was correct in that respect of the Japanese defenders on Corregidor but it was only a guess.

Information about the defending Japanese was scant and was a product of guesswork as the defenders were dug in deep. The staff estimates placed about 850 defenders on the island but in fact there were over 5,000 Japanese. Commanded by CPT Akira Itagaki of the Imperial Navy, they were organized into provisional units and assigned sectors of the coastline to defend. The defenders made use of both caves and tunnels that went deep underground on the island. Over half of the garrison was positioned on Malinta Hill in reserve with detachments in the scattered ravines around the island. Fortunately for the Americans, this left the open ground of Topside fairly lightly defended from air assault. Although warned of a possible airborne assault CPT Itagaki conducted a careful terrain analysis and thought a parachute attack was not feasible. Consequently, he made no preparations for one. The Americans received a report that the Japanese had erected sharp poles and other anti-parachute obstacles on the proposed drop zones but it was untrue. Quite simply, the Japanese thought an airborne assault was impossible and made no plans to defend against one. However, Itagaki and his men were determined to hold Corregidor to the last man.

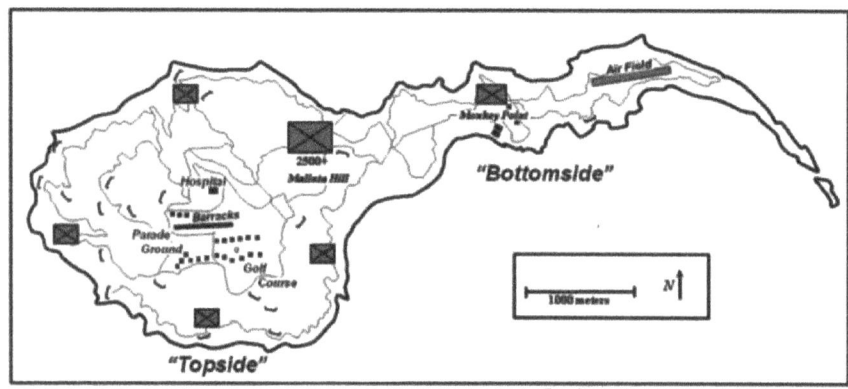

Map 1. Japanese dispositions on Corregidor, January 1945.

COL George M. Jones commanded the 503d PRCT during the operation and received the mission from XI Corps on 3 February. He and his staff were tasked to do the specific planning for the airborne operation, although no doubt the XI Corps, Army Air Force, and US Navy staffs were involved. COL Jones arranged for an aerial reconnoiter of Corregidor for himself, the battalion commanders, and selected staff officers. Just before the operation, the regimental commander held a formation and briefed each member of the 503d PRCT on the mission and concept of the operation. Each man knew exactly what he was to do and what was expected of him. The pilots of the 317th Troop Carrier Group attended every operations briefing and were encouraged to provide comments and suggestions, of which many were incorporated into the plan. Liaison teams from the 7th Fleet and the 5th and 13th Air Forces were present and representatives of the 3-34 IN were on hand as well. Communications and resupply issues were identified and provided for.

Topside was the key terrain feature on Corregidor and dominated the island. The airborne planners weighed the options and decided to designate the old parade ground and the nearby small golf course as the drop zone. Each of these was only approximately 300 meters long by 200 meters wide. In addition to their small size, the area in and around the drop zone was interspersed with the ruins of the old barracks and post buildings as well as the shattered trees and shell craters from the preparatory bombing. Correlating the factors of wind speed and drift, the 51 available C-47 transports were required fly multiple passes at 400 feet above the drop zone for over an hour to drop each lift. This was because the very short time over target allowed only six to eight paratroopers to exit on each pass. Little anti-aircraft fire was expected as the bombing and strafing which began on 22 January would keep the Japanese down during the drop but the first lift jump casualties were expected to be about 20 percent due to the condition of the fields and expected high winds. Incidentally, any excessive drift in the 25 knot winds would send paratroopers over the island and into the sea incurring even more losses. The final plan had the 3d Battalion making the morning drop at 0830 and securing the drop zone. There it would prepare for the 2d Battalion's arrival at 1230. The 1st Battalion was slated for arrival the next morning. The amphibious force was scheduled to hit the beach at 1030 to take advantage of any covering fire the airborne forces were able to provide. After these events, the plan was extremely broad, devoid of details largely because the precise location of enemy forces was unknown. The regimental commander divided Topside into battalion sectors and charged his units to locate the Japanese and kill them. The regimental order, for example, directed 2d Battalion to "exploit the area to the north and west of the drop area destroying all enemy encountered." Once Topside was cleared of enemy forces the combined airborne and infantry forces were to sweep eastward to clear the rest of the island. Air and naval forces would support the operation by fire throughout.

Inherent with any airborne operation is the probability that unit cohesion would be highly difficult to maintain as paratroopers get scattered across a hostile drop zone. Each man of the 503d PRCT was trained to link up with any friendly unit he came across and continue the mission if he was unable to find his own element. Each man knew the concept of the operation and the commander's intent. Commanders at all echelons were provided with radios for communicating with their higher headquarters but terrain and other factors could render these useless. Officers and noncommissioned officers were expected to take the initiative in these situations and continue on with the objective. The commanders of the 503d PRCT and the 3-34 IN knew their men well and put high trust in their subordinate commanders' judgment.

Preparatory fire began in earnest as the date for the assault drew near. By 16 February, the 5th and 13th Army aircraft had dropped approximately 3,000 tons of bombs and napalm on Corregidor. Known and suspected gun emplacements were hit and strafed. The naval bombardment had begun on 13 February with the five heavy cruisers, five light cruisers, and 14 destroyers of Task Group 77-3 directing most of their fire at Topside. The Navy also positioned PT boats to rescue any paratroopers or infantrymen that found themselves in the water.

The airborne phase began on schedule on the morning of 16 February. There was no opposition during the first pass and only sporadic firing from the Japanese during the first lift. By 0945, 3d Battalion was on the ground. COL Jones and much of his headquarters landed and assembled as well. The 3d Battalion immediately went to work establishing a perimeter for the inbound second lift as well as preparing to clear Topside and provide cover fire for the arrival of the 3-34 IN by sea. The arriving infantrymen were aboard 25 landing craft medium (LCM) under the watchful eye of two .50 caliber machine guns from the 503d PRCT covering their approach.

The men of the 3-34 IN arrived at the beach on schedule at 1030 and the first four waves met no opposition. Japanese machine guns opened up though as the fifth wave landed and detonated prepositioned mines. A supporting M4 Sherman tank, an M7 self-propelled gun, and a 37mm anti-tank gun were lost. However two companies of the 3-34 IN pushed on and were atop Malinta Hill by 1100. It was apparent that both the air drops and amphibious landings were a complete surprise to the Japanese defenders. The naval and air bombardment had kept the defenders under cover and the coordinated assaults diverted attention from each other. By the time the Japanese had recovered their shock, the first objectives of the operation were held and the Americans were firmly on the island.

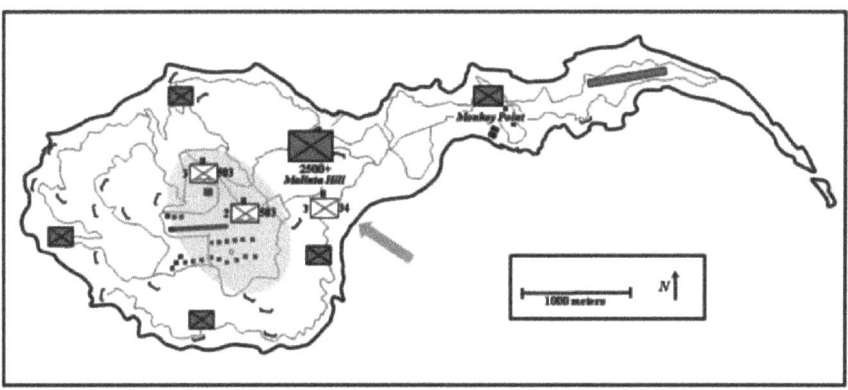

Map 2. The initial assault, 16 February 1945.

The first lift of the 3d Battalion exceeded the casualty estimates of 20 percent. Alarmed at these losses, there was a staff discussion to consider halting the second lift but COL Jones decided to continue. The 2d Battalion arrived at 1240, almost a half hour behind schedule. This drop also came under sporadic Japanese fire but suffered fewer casualties than the first. The 2d Battalion relieved the 3d Battalion in perimeter defense. Meanwhile, the 3d Battalion was tasked to search and destroy all remaining Japanese forces on Topside. By nightfall, most of the old American facilities around the parade ground were secure. There were Japanese defenders in the many ravines along the coast but no one knew exactly where or in what strength.

The combined first day regimental casualty rate was 14 percent, most of which was caused by injuries on landing. There were only 55 combat casualties, a number far less than anticipated. That evening COL Jones requested that XI Corps cancel the morning drop and send the remaining battalion by sea. The request was quickly approved and the 1st Battalion, 503d PRCT marshaled to the landing craft. This battalion task force entered the fight roughly on schedule and avoided the injuries of an airdrop in marginal weather and with a drop zone in poor condition.

The Japanese were in no mood to surrender but they were badly outclassed in firepower and outmaneuvered. They also suffered a loss that morning which proved catastrophic. During the first lift, a small number of American paratroopers that drifted away from the drop zone landed near an observation post being visited by Captain Itagaki, the senior Japanese commander on the island. Quickly forming an ad hoc squad, these Soldiers killed Itagaki and apparently there was no clear line of command succession. Even if there had been, the wire communications relied upon by the garrison were cut by the days of bombing and gunfire. The only means for the Japanese to communicate was by courier which was nearly impossible in the situation. For the defenders, the battle quickly devolved into a series of badly coordinated "banzai" charges and small unit actions. It was fanaticism at its best but it had no real hope of repelling the Americans. On the first day alone, the Japanese lost over 300 men and the second day would see the loss of nearly a thousand more for little or no gain.

The 1st Battalion and other reinforcements landed ashore in the afternoon of 17 February and joined in the efforts to clear Topside, an effort which would take six days. The Japanese fought ferociously throughout. On the night of 21 February, they detonated the tons of ammunition and explosives stored in the tunnels under Malinta Hill, presumably as a prelude to a counterattack westward or a withdrawal to the east. The massive blast literally shook the entire hill and hundreds of Japanese perished inside the tunnels. Several hundred of them did make it to Bottomside while about 600 massed for a counterattack to the west. Heavy indirect fire was brought against this force and it too retreated eastward after suffering tremendous casualties.

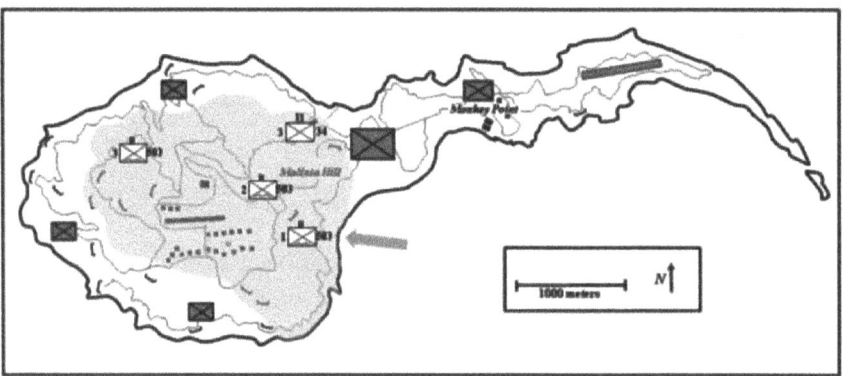

Map 3. The second day, February 17, 1945.

After the three parachute battalions were on Topside by Day 2, the operation plan directed each unit to operate in a sector with 1st Battalion on the south and southeast side, 2d Battalion on the east, and 3d Battalion on the west. COL Jones had not designated specific objectives but gave each subordinate commander the latitude to send forces where he expected to find enemy units, the overarching objective being the complete clearance of the island. The plan only required the battalions to coordinate their movements and actions with higher headquarters and each other. During the operations on Topside, a general pattern of combat developed. First, aircraft or naval fire support were called upon to strike known or suspected Japanese positions. The Americans then assaulted the position immediately as the fires lifted. If this failed, the 75mm pack howitzers or one of the few supporting tanks was brought forward for direct fire. If that failed, small teams armed with flamethrowers and demolitions crept forward to seal cave or tunnel entrances. The platoon leader of the regimental Demolition Platoon developed a method to neutralize the larger fortifications that proved impervious to these methods.

All of these clearing operations placed a premium on personal initiative and decentralized control to the lowest levels. In fact, several units innovated by taking advantage of the Japanese desire to regain positions they had lost to the Americans despite the risks associated with the action. During the day, the US units used fire and maneuver to dislodge Japanese elements from their bunker or machinegun positions. Just before dusk the paratroopers abandoned the position but only after ensuring that it was targeted by indirect fire and nearby crew-served weapons. The Japanese reoccupied the position after dark. Once the sun came up, they were easily destroyed by the fires planned the night before.

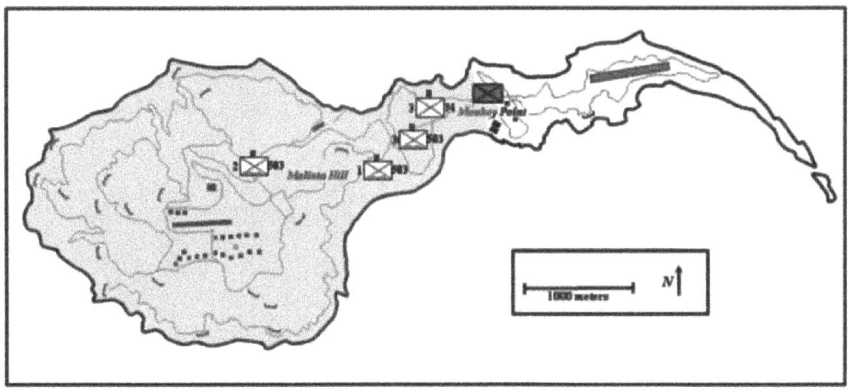

Map 4. The final push, 18-26 February 1945.

With Topside generally secured by 24 February, the US force pushed eastward to clear and secure Bottomside. There continued to be stiff resistance and suicidal counterattacks but the end was in sight. Organized resistance ceased shortly after 1100 on 26 February when the Japanese detonated an underground arsenal at Monkey Point, killing most of the few remaining defenders. COL Jones formally presented the island of Corregidor to GEN MacArthur on 2 March and the battle was officially over. Both the 503d PRCT and 3-34 IN were quickly withdrawn from the island to prepare for operations in the south Philippines.

For Further Reading

Gerard M. Devlin. *Back to Corregidor: America Retakes the Rock*. New York, St. Martin's Press, 1992.

E.M. Flannagan Jr. *Corregidor, The Rock Force Assault*, 1945. New York, Random House Digital Collection, 2012.

Samuel E. Morrison. *The Liberation of the Philippines: Luzon, Mindanao, the Visayas 1944-1945* (History of United States Naval Operations in World War II), New York, Castle Street Press, 2001.

The Six Principles of Mission Command

1. Build Cohesive Teams through Mutual Trust
2. Create Shared Understanding
3. Provide a Clear Commander's Intent
4. Exercise Disciplined Initiative
5. Use Mission Orders
6. Accept Prudent Risk

Mission Command in the Corregidor case

1. Build Cohesive Teams Through Mutual Trust. All units of this operation were veteran formations, most of which had worked together in the past. Cohesive teams and mutual trust had been formed through hard training and combat. Although a large number of replacement personnel arrived just days before the battle, the unit command structures were firmly in place. Combat leaders at all echelons were well trained and experienced. At the task force level the 3-34 Infantry Battalion was selected in large part because it had worked with the 503d PRCT in the past. Additionally, the air support and air transport elements also had a long relationship with the 503d PRCT.

2. Create Shared Understanding. All participants and supporting elements had a shared understanding the mission and commanders intent. The importance of the recapture of Corregidor was clear to all. All of the men were briefed on each phase of the operation and their part in it. Pilots and staff of the air and naval forces supporting the operations attended the planning sessions and provided input. These forces also insured they had common radios and procedures in place, and provided liaison teams as well.

3. Provide a Clear Commander's Intent. The commander of 503d PRCT provided a clear commander's intent to subordinate commanders and down to each man. Although the concept of seizing and clearing the island of every Japanese defender is a simple one, the phases of the operation were highly choreographed and each element knew what was expected of it. Each paratrooper and infantryman on Corregidor and those supporting them from the air or sea was made aware of the plan and proved adept a reacting to a fluid situation. In the regimental operation order, mission statements to the subordinate battalions were general and clear. One example of this type of mission statement was that received by 2d Battalion, "exploit the area to the north and west of the drop area

destroying all enemy encountered." Battalions were given sectors to clear rather than specific objectives. Coordination was required with higher headquarters and lateral units but freedom of action was maximized.

4. Exercise Disciplined Initiative. The paratroopers and soldiers who made the landings were veterans of the Pacific war, and were highly adept at the squad and platoon operations used to clear bunkers and fortifications held by determined Japanese defenders. This disciplined initiative was key to the success of this operation as small units rooted out the Japanese defenders scattered across the island. Both paratroopers and infantry excelled at this as it had become routine in combat operations in the Pacific. High leader casualties and poor communications demanded that all combat commanders took stock in the situation and acted accordingly towards mission accomplishment. On Corregidor this was displayed repeatedly by squad and platoon leaders. One excellent example on Day 1 of the operation was the creation of the ad hoc squad that killed the Japanese commander. At the task force level COL Jones made the decision to cancel the third drop and bring the battalion in by sea. His chain of command concurred with this assessment from the man on the ground and supported the decision fully. At battalion-level and below, units had freedom of action to introduce new techniques and tactics. The innovation of abandoning positions to Japanese at night is one good example of this. This innovation, which ran counter to accepted practice of holding seized terrain, allowed for the relatively easy elimination of the enemy once morning came.

5. Use Mission Orders. The 503d PRCT used the formal orders process prior to the drop. The nature of airborne operations during the war necessitated planning for a wide dispersal of soldiers during the drop and difficulty in organizing into original formations once on the ground. The mission orders for this operation took this all into account. Paratroopers were expected to carry on with the mission even if their leader was not present and to use whatever means at hand to do it. The orders were broad and maximized freedom of action for subordinate units. Battalions were assigned sectors rather than specific objectives and timetables. At lower levels, small units conducted patrols and once engaged, fire and maneuver, developing their own innovative techniques and tactics to solve problems such as those posed by Japanese elements inside reinforced tunnels.

6. Accept Prudent Risk. The use of an airborne drop was intended to surprise the Japanese and avoid the high casualties of an amphibious assault. The Americans accepted prudent risk by attempting an airborne assault on Corregidor. The drop zones were very small, the terrain hazardous,

the winds high, and the number of Japanese badly miscalculated. If the assault went badly there would be great difficulty in extracting the force, but the odds for success were calculated and the mission went forward. However, one can wonder if the operation would have gone forward as planned had the Americans known there were over 5,000 Japanese on the island. During the operation COL Jones' decision to call off the third drop proved very prudent. He calculated the anticipated losses due to injury as too great and instead brought the battalion in by sea. His decision proved correct as the injuries were avoided and the battalion arrived intact only a few hours behind schedule.

Assault River Crossing at Nijmegen, 1944

Donald P. Wright, Ph. D.

In August 1944, the Allies were pushing toward Germany in attempt to defeat the Third Reich before winter arrived. Standing between Allied forces and the German heartland was the Rhine River, which Hitler planned to use to use as a formidable line of defense. Understanding that crossing the Rhine would take time and cost many lives, the Allied Command planned an audacious operation called Market-Garden that would quickly seize a major bridge over the Rhine in the Netherlands. Once secure, that bridge, located in the Dutch town of Arnhem, would be used as a gate through which Allied forces would pour into Germany. Success in this operation would require surprise and speed. To gain surprise, the Allied Commanders chose to drop two US Airborne Divisions, the 82d and 101st, in the Netherlands to seize and secure a series of six bridges along the road to Arnhem. The bridge over the Rhine itself would be seized by the British 1st Airborne Division dropped near the town of Arnhem. Once the Airborne forces were in control of the bridges, the British XXX Corps, a powerful force composed of mobile armored units, would fight quickly up the route to relieve the British Paratroopers in Arnhem and secure the gateway into the Third Reich.

The 82d Airborne Division's mission was to capture key terrain in the vicinity of the Dutch cities of Grave and Nijmegen. This included five bridges, the largest of which spanned the Waal River in Nijmegen. Because there was a limited number of aircraft available to drop the paratroopers and tow the gliders, the division's combat power would land in the Nijmegen area over a three day period. So MG James Gavin, the division commander, designated the main Nijmegen Bridge as a priority, tasking the 508th Parachute Infantry Regiment (PIR) to send a battalion to seize that bridge as soon as possible on the first day of the operation. A railroad bridge over the Waal downstream from the main Nijmegen Bridge was not a priority objective. Two other regiments, the 504th PIR and 505th PIR, were directed to seize and hold four bridges in the vicinity of Grave and high ground near the town of Groesbeek on the first day as well. Glider-borne forces, including artillery and support units, would follow on the second and third days to help consolidate the gains made by the paratroop regiments.

Figure 1. Plan for Market-Garden.

The jumps on the first day, 17 September, went well with little initial German opposition. Most of the division's first day objectives were seized quickly but the main Nijmegen Bridge remained in German hands. An assault by Company A, 508th PIR had run into staunch German resistance on the south side of the bridge. Two additional American assaults on the bridge on Day 2 came within a block of the bridge entrance but were ultimately repulsed as the Germans had greatly reinforced their positions.

The problem for MG Gavin and the 82d Airborne was how to secure the bridge so that the tanks of XXX Corps, rapidly approaching from the south, could cross the Waal and make their way to Arnhem to relieve the 1st Airborne that had seized the bridge over the Rhine and were holding on to it by their fingernails. On Day 2 of Market-Garden, Gavin began thinking about the tactical problem posed by the strong German positions on the south side of the bridge but other priorities prevented him from launching an immediate attempt to seize it. On Day 3 when reconnaissance elements of XXX Corps made contact with the 82d, Gavin was forced to act and finalized a wholly new plan that seemed to be the only means of meeting the intent of the larger operation. Although not equipped with assault boats, Gavin intended to envelop the German positions on the bridge by sending two battalions of the 504th PIR across the Waal River in a variety of civilian watercraft. Once on the north side of the river, the battalion's Soldiers would attack and seize the northern end of the Nijmegen Bridge. At the same time, 2d Battalion, 505th PIR, with support from a British tank battalion, would attack the southern side of the bridge. Gavin's hope was that the simultaneous attacks on both sides would force the Germans to retreat, leaving the bridge open to the Allies.

When a quick search turned up few civilian boats, British staff officers in XXX Corps arranged for their engineers to bring assault boats up to Nijmegen for the crossing but because the boats could not be at Nijmegen until the afternoon of the next day (Day 4), Gavin unhappily planned for the assault crossing to begin in the afternoon. To mitigate the risk of a daylight crossing, he arranged for a great deal of fire support, including mortars, tanks, artillery, and rocket-firing Typhoon aircraft, targeting the far side of the Waal River which was defended in strength by German forces. Gavin briefed the entire plan to the XXX Corps staff and the leaders of the 504th PIR on the evening of Day 3.

The assault crossing would be led by 3d Battalion, 504th PIR, commanded by a 27 year old MAJ Julian Cook. Cook had served with the regiment since Sicily and had rigorously trained his battalion, made up of hardened veterans, in England before the 504th PIR deployed to Europe. Once Cook got his rifle companies across and secured a bridgehead on the north side of the river, the 1st Battalion of the regiment would follow and secure the western flank. The landing site was approximately two miles west (down river) from the Nijmegen Bridge. After consolidating on the northern bank, two of Cook's companies (H and I) would move east down the river bank, locate an earthen railroad embankment, and follow that north until they hit the road leading from the main Nijmegen Bridge. They would then turn southeast and assault the north end of the bridge moving companies abreast, one on either side of the road. Company G would follow to protect the rear of the two companies in the assault. By early evening, Gavin hoped to have the bridge in allied possession and the tanks of XXX Corps rolling across it on their way north to Arnhem.

The realities of the terrain and the enemy's dispositions posed serious obstacles to the operation achieving a quick victory. When MAJ Cook first saw the intended crossing site, he realized for the first time that the river was 400 yards wide and its current was swift. At that point, the battalion commander recalled thinking that someone above him had come up with "a real nightmare." He then saw that if they succeeded in getting across this watery expanse, his Paratroopers would then have to cross a flat plain devoid of cover and concealment and which was 700 hundred yards in length until they could finally find cover behind a 30 foot high dike. Cook and his staff officers quickly identified German gun positions along the northern bank that could sweep the river and plain with machinegun and cannon fire. Several Dutch stone forts on the north side served as strongpoints for the German defense of the bridge at Nijmegen and would have to be attacked if paratroopers were to make it all the way to their objective. Finally, there was the railroad bridge on the river approximately 1,500 yards to the east of the crossing site. German units had set up 20 mm gun positions on that structure that could easily fire down on the men of the 504th PIR as they crossed the river and plain.

Despite his concerns, Cook planned for his forces to consolidate at the dike and then follow the scheme of maneuver that directed H and I Companies to assault the north end of the bridge by moving southeast down the road. All understood that the intent of the division commander was the seizure of the northern end and the opening of the bridge. In the early afternoon on the day of the assault, officers briefed their men on the mission and intent as they waited for the boats to arrive. Many recalled feeling that the operation was like a Normandy-style landing and that they had not trained for that type of mission but the Soldiers also understood that the Nijmegen Bridge had to be taken if the British Paratroopers at Arnhem were to be relieved.

The 26 boats arrived at the crossing site at 1430, approximately 30 minutes before the close air support would arrive and artillery barrage would begin. The Soldiers were surprised to find that they were small craft (19 feet long) with a wood frame and canvas skin. A US Engineer company had been assigned to operate the boats and found quickly that many of them were missing paddles. The Engineers went ahead and assembled them, after which the units in the first wave – Companies H and I and part of the battalion HQ - moved to their assigned boats and began loading equipment and ammunition. To many, it was clear that the boats would have a hard time making it across the Waal even without the Germans shooting at them but there was little time to ponder their plight as the artillery began to fire and the Typhoons arrived to pound the German positions on the far side of the river. Smoke rounds quickly formed a screen that would provide some concealment for the Soldiers.

For most of the paratroopers, getting the boats into the water and moving across the river was a terrifying experience. Despite the smoke screen, enemy gunners quickly discovered the activities at the launch site and began firing at the men struggling with the heavy boats. Once on the water, men paddled with whatever they had to include paddles, rifle butts, and hands. German machine gun and mortar fire hit many of the boats during the crossing. The current made some of the boats almost impossible to steer. MAJ Cook, the battalion commander, led the first wave and recalled chanting, "Hail Mary, Full of Grace" as he paddled.

Of the 26 boats that left the southern bank, only half made it across in usable condition. Some did not make it at all. Officers and NCOs who made it to the north side quickly rallied groups of paratroopers that were still alive and not severely wounded and began leading them across the plain through more German fire. The wounded were gathered at a makeshift aid station. The Engineers began paddling the usable boats back to the southern side of the river. They would ultimately make several trips across the river, bringing the remainder of Cook's battalion over as well as elements of the 1-504 PIR.

Those in the first wave that made it to the dike quickly organized, located enemy positions on the dike, and began a ferocious battle for control of that key terrain. Many Germans surrendered while others had to be killed with grenades and in brutal hand to hand combat. The chaos during the river crossing and sprint to the dike had broken up squad, platoon, and company integrity. The paratroopers at the dike instead formed small groups and, understanding the mission and intent, had taken control of that position and begun to consolidate.

The battalion's disorganization meant that MAJ Cook's plan to have H and I Companies attack abreast down the road toward the north end of the bridge was no longer feasible. Instead, officers and NCOs formed small groups and moved toward the Nijmegen Bridge, their ultimate objective. One of Company I's Soldiers, SGT George Leoleis, recalled the actions of his small group, stating, "We were separated from any other men but we knew in what direction to head for, down the road toward the bridge." The commander of Company G found that by late afternoon, the group he led included Soldiers from companies H and I as well as his own company and the battalion communications and medical sections. MAJ Cook, and his operations officer, CPT Keep, quickly put together a group of 30 men and began moving east from the dike through orchards and down ditches. Keep recalled that they formed ad hoc squads and used bounding movements across open areas and from one house to another as they approached the bridge. By quickly grabbing the initiative in this manner, Keep believed they were able to keep the German defenders off balance, preventing them from reorganizing.

1LT Jim Megallas, a platoon leader in Company H, gathered about a dozen men from his platoon and moved to assault one of the Dutch forts from which the Germans were using a 20 mm gun to fire at the dike and at units crossing the river. Megallas' force concentrated small arms fire on the fort, suppressing the German gunners. One of Megallas' NCOs, SGT Leroy Richmond, then swam the moat surrounding the fort and tried to kill the Germans inside. Megallas quickly called him off, and remembering that the bridge was the objective, decided to move his group further east, leaving the fort for other units to seize.

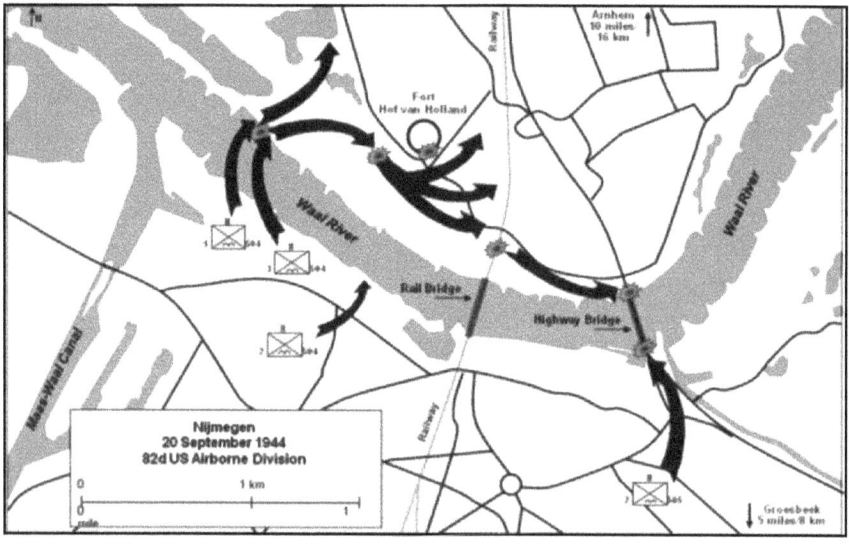

Figure 2. Assault river crossing at Nijmegen, 20 September 1944.

Some small groups followed the battalion plan and moved north along the railroad embankment to the road. There they met a great deal of German resistance. Another group led by CPT Carl Kappel, the commander of H Company, reached the embankment and rather than go north according to the plan, turned south toward the river. They hoped to find a way under the railroad bridge that would then open up a direct route to the main Nijmegen Bridge just 1,000 yards away. Kappel's group was so successful that it seized the railroad bridge from the Germans. CPT Moffatt Burriss, the commander of I Company, then took charge of another ad hoc group gathered at the railroad bridge and began moving east toward the main bridge. Along the way, they had had to stop and clear a number of buildings on the banks of the river. Burriss' group arrived under the main bridge at the same time that British tanks started crossing the bridge from the south side where the attack of the 505 PIR had been successful. The paratroopers from 3-504 went running up steps leading to the road surface above and met relatively little German resistance. In fact, the Germans defending the main bridge, threatened by the envelopment from the river crossing, had begun to pull back to the north away from the bridge. By 1915 that evening, the intact bridge was in Allied hands.

The fight at Nijmegen had been a success. Part of that victory can clearly be attributed to MG Gavin's vision of using an assault river crossing to envelop the Nijmegen Bridge from two directions but equally important was the way in which the Soldiers of the 3-504 PIR carried out the assault. Without their courage, devotion to the mission, and initiative at individual- and small-unit levels, it is difficult to envision how the crossing operation could have succeeded. The cost in lives was high. Twenty eight paratroopers from the 3-504 PIR made the ultimate sacrifice. H Company lost 15 killed or missing in action and suffered another 38 wounded. Another 40 of the battalion's Soldiers were wounded but the actions that day allowed Allied forces to move toward Arnhem. Although the bridge over the Rhine did not remain in British hands, XXX Corps was able to extricate part of the surrounded 1st Airborne Division, saving thousands of British Soldiers from death or capture.

For Further Reading

Cornelius Ryan, *A Bridge Too Far* Simon Schuster, New York, 1974.

Phil Nordyke, *All American All the Way: The Combat History of the 82d Airborne Division in World War II*, Zenith Press, Saint Paul, Minnesota (2005)

T. Moffatt Burriss, *Strike and Hold*, Potomac Books, Dulles, Virginia, 2000.

The Six Principles of Mission Command

1. Build Cohesive Teams through Mutual Trust
2. Create Shared Understanding
3. Provide a Clear Commander's Intent
4. Exercise Disciplined Initiative
5. Use Mission Orders
6. Accept Prudent Risk

Mission Command in the Nijmegen case

1. Build Cohesive Teams through Mutual Trust. Many of the 3-504 PIR's Soldiers had served together in combat for years. The battalion commander, MAJ Cook, had been with the regiment since 1943. He had trained his men hard while in England in 1944 preparing for operations in Europe. CPT Moffatt Burriss, the H Company Commander, had been with the 3-504 PIR throughout this period as well and commanded HQ Company during the Anzio invasion. By the time they were preparing for the river crossing, the men of the 3-504 had fought together for days, learning each other's strengths and weaknesses, building trust, and developing greater cohesion. The battalion commander and his company commanders likewise had established close relationships.

2. Create Shared Understanding. MG Gavin briefed his plan for the assault crossing to his entire staff and 504th PIR leaders. MAJ Cook was then able to brief his leaders before the assault started. While waiting for the boats to arrive, subordinate leaders explained the mission to their Soldiers. All understood that the overarching mission was to seize the northern end of the main Nijmegen Bridge.

3. Provide a Clear Commander's Intent. From Gavin's level down to squad leaders in the 3-504 PIR, it is evident that commander's intent was effectively passed down. At Gavin's level, he understood that seizing the Nijmegen Bridge was critical to the success of Market Garden. At the battalion level and lower, the actions of the small groups on the northern side of the river are evidence of an intent understood by all.

4. Exercise Disciplined Initiative. At Gavin's level, the division commander and his staff took the initiative to develop the plan for an envelopment of the bridge by using an assault river crossing. The actions of the small groups on the northern side of the river show commissioned officers and NCOs, in the chaos of combat, organizing small groups to

move toward the objective. SGT Leoleis statement about his small group taking initiative to achieve the mission is an excellent example of this: "We were separated from any other men, but we knew in what direction to head for, down the road toward the bridge." 1LT Megallas' decision to end his attack on the Dutch fort in order to move his troops toward the most important objective is an excellent example of initiative that was disciplined. Because he understood the commander's intent and needs of the mission, he chose to bypass the fort, despite the fact that it represented a very real threat to US units.

5. Use Mission Orders. As noted above, using general briefings and discussions before the river crossing, the mission was made clear to all down to Soldier level. The scheme of maneuver was simple and the objectives very clear. The mission orders were the key to success. They allowed for the small ad hoc groups to ignore the planned approach to the bridge and retain freedom of movement and decision-making. This enabled them to get to the northern end of the bridge in an unplanned but effective way.

6. Accept Prudent Risk. MG Gavin had been managing risk carefully since his division landed outside of Nijmegen. His attempts to mitigate risk had led to his decision not to risk his troops in an all out frontal assault on the southern end of the bridge but by Day 3, the larger objectives of Market Garden, specifically the relief of the British 1st Airborne, overrode these concerns and Gavin made the decision to make the river crossing, understanding the danger involved. He attempted to mitigate the risk inherent in the daylight crossing of a river against an entrenched enemy by arranging for fire support and a smoke screen but this was only partially successful.

Sicily, 1943

Initiative Prevails at Biazza Ridge

Gregory S. Hospodor, Ph.D.

The largest engagement involving paratroopers of the 82d Airborne Division during the July 1943 invasion of Sicily, codenamed HUSKY, occurred on what the All-Americans called Biazza Ridge. The intense fighting was not part of the detailed airborne pre-invasion plan. Nevertheless, this "accidental" battle between a small primarily paratroop force led by COL James Gavin and powerful elements of the German *Hermann Goering* Division played a key role in securing the success of the landings. Furthermore, the battle for Biazza Ridge and the chain of events leading up to it illustrate the importance of flexibility, resourcefulness, personal initiative, decentralized command, cohesiveness built through rigorous training, and reliance on the commander's intent.

The road to Biazza Ridge began in January 1943 at a conference held in Casablanca, French Morocco. In a series of meetings, strategic-level civilian and military leaders from the United States and Great Britain decided to follow up the conquest of North Africa by invading Sicily. They hoped that taking the three-cornered, roughly 10,000-square-mile island would achieve three ends: secure sea lines of communication through the Mediterranean thereby freeing up a significant amount of indispensible shipping, knock Italy out of the war, and relieve pressure on the Russian front by drawing German strength south.

It fell to GEN Dwight D. Eisenhower, the commander of Allied forces in the Mediterranean, to turn strategic intent into a workable operational plan while at the same time seeing the North Africa campaign to a successful conclusion. The Allied order of battle for the invasion included the equivalent of over 12 American, Canadian, and British infantry, armored, and airborne divisions; over 4,900 aircraft of the US Army Air Force and the Royal Air Force; and over 3,500 ships under the control of the US Navy and the Royal Navy. Eisenhower's orders from the Combined Chiefs of Staff required that a separate headquarters be set up to plan the ground portion of the invasion. Consequently, Force 141, which later became 15th Army Group Headquarters commanded by Field Marshal Sir Harold Alexander, began planning in January 1943. Subordinate headquarters for each national ground contingent, Force 545, later 8th (GB) Army commanded by General Sir Bernard Montgomery, and Force 343, later 7th (US) Army commanded by LTG George Patton, were also stood up. After considerable debate, no little acrimony, and a lot of staff work, the final plan necessitated that, according to 7th Army Field Order Number One, "FORCE 141 ... supported by combined US and British Air and Naval Forces, assaults the southeastern portion of SICILY to capture it as a base for further operation." Thus, Patton and Monty's forces would be landing side by side on an over 100-mile front with the mission of establishing a firm lodgment ashore.

Patton's lineup for the impending invasion along a 70-mile stretch of coast included II Corps Headquarters under LTG Omar Bradley, four infantry divisions (1st, 3rd, 9th, and 45th), one armored division (2d) and the 82d Airborne Division. The final plan envisioned 3d Division (reinforced) landing furthest west with the missions of seizing the small port of Licata and providing a firm flank for 7th Army. Elements of the veteran 1st Division and two ranger battalions would land in the middle at Gela, seize the town and the Ponte Olivo airfield, and serve as a floating reserve. It was here that the relatively flat open terrain invited a German and Italian counterattack. The 45th Division, a National Guard unit fresh from the United States, would land furthest east, link in with the 8th Army, capture the Cosimo and Biscari (today's Acate) airfields, and exploit inland. Elements of the 2d Armored Division would land with the assault waves and serve as floating reserve. The 9th Division, with 39th Regimental Combat Team and artillery on call, remained in Tunisia as the Army's follow-on reserve. The greatest threat to the US operation, and indeed any amphibious landing, was that the troops storming ashore suffer a coordinated counterattack early in the assault phase. To mitigate this risk, especially apparent in the vulnerable Gela sector, planners quickly settled upon the use of airborne troops. Broadly stated, the mission of the paratroopers was to assist the amphibious troops to get and stay ashore. Planners quickly tapped the 82d Airborne Division for the job.

The concept of divisional airborne assault was relatively new to the United States Army. Indeed, many of the leaders of the 82d Airborne, such as its commander, MG Matthew Ridgeway and the 505th Parachute Infantry Regiment's commander, COL James Gavin, played a key role in turning the idea into reality. The Army activated the 82d as the first parachute division in March 1942. Like many units in the rapidly growing army, it suffered growing pains, perhaps more than most because of its specialized role. There were shortages of parachutes, transport aircraft, and gliders. Frequent reorganization and personnel reassignment meant that training and unit cohesion suffered. Consequently, a March 1943 inspection revealed that the unit had completed one-third the amount of training of a regular infantry division and was thus unprepared for combat operations. There was, however, no shortage of volunteers because of the allure of extra pay, jump boots, a unique uniform, and the promise of rigorous training and adventure. Nor was there a shortage of leadership because the unit tended to attract and promote leaders for whom the challenge of starting something new was seen as an opportunity and who relished an environment that prized the exercise of individual initiative more than many conventional army units. Generally, then, the quality of the division's human material was a cut above its contemporaries. When the 82d arrived in North Africa in early May 1943, two months before the first test of the airborne division concept in battle, it remained a unit with vast but as yet unrealized potential.

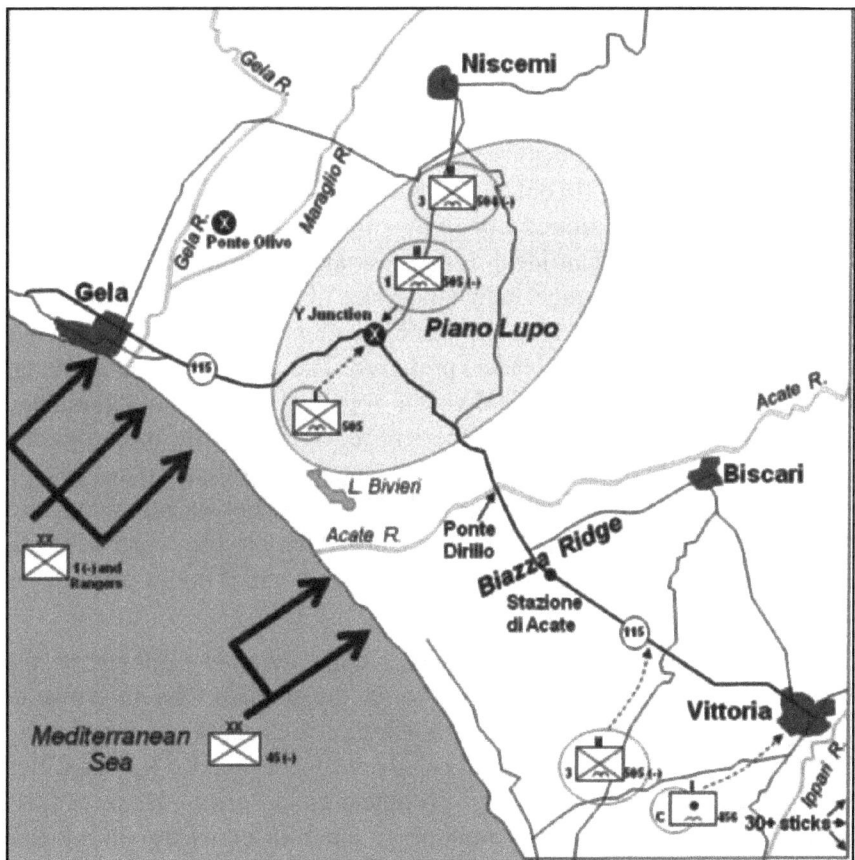

Figure 1. 82d Airborne Division actual drops on Sicily.

In the blast furnace of North Africa, leaders honed the edge of the division and pushed their troopers hard. At Oujda, French Morocco, the daytime heat was intense and training typically began at dusk and continued until dawn. Infantry tactics and night navigation were practiced over and over again. Training emphasized individual initiative and flexibility because the unique nature of airborne delivery almost guaranteed the necessity of *ad hoc* reorganization on the ground, especially at night. Furthermore, after landing, troopers could not count on immediate resupply and were thus familiarized with enemy weapons and ammunition as well as encouraged to innovate with whatever they found in order to accomplish the mission. When possible, training was tailored to the requirements of the mission while keeping the objective of the invasion a secret. Paratroopers could expect to face at least some of the concrete and fortified positions that dotted the invasion zone. Accordingly, leaders constructed mock-ups of characteristic positions and conducted live fire exercises in their reduction by day and night. The importance of cutting telephone and telegraph lines was also stressed and physical training reached a new pitch. After the 82d moved to the airfields in Tunisia from which it would launch the assault, leaders went over the plan in detail with the help of aerial photographs and maps. One benefit of the deployment to North Africa was that the combing of experienced men to fill out new airborne units ceased, which along with the rigorous training, proved a boon to unit cohesion. Conditions for drop training proved less than ideal as too few aircraft, a divided air-ground command structure, frequent high winds, lack of suitable "soft" drop zones, and the necessity to stage forward to embarkation airfields in Tunisia on 21 June led to less jumps than leaders preferred. The 52d Troop Carrier Wing, the primary lift assigned to the division, was qualified for both glider and parachute operations but was inexperienced. Consequently, drop training, especially at night, proved less than satisfactory and often resulted in high injury rates. Nevertheless, the 82d, with the clock ticking rapidly toward D-Day, maximized its training time in North Africa to the extent possible. It was as prepared as it could be for the test to come if not as prepared as its leaders wished.

As unit leaders honed the division's edge, detailed planning for the drop went forward. The primary mission of the division within the overall Husky plan was to assist the troops landing on the beaches by interdicting Axis reserves. Initial planning conducted outside the division had identified the Gela area as most vulnerable to counterattack. Because the open Gela plain was unsuitable for light infantry to confront conventional infantry and armor, planners ascertained the relatively treeless high ground to the east of the plain, called the Piano Lupo, was appropriate for the drop. On the Piano Lupo, the terrain was more restrictive, which would benefit the paratroopers as they battled heavier enemy forces. Three further considerations also favored the site: it possessed suitable drop zones;

through it ran the best, and thus most likely, avenues of approach to the 1st Division and Ranger landing beaches; and taking the high ground would deny direct observation of the landing area and reverse slope gun positions to enemy artillery. Army group planners envisioned that, once the 1st Division was established ashore and linked up with the paratroopers, the 82d would assist in taking the important Ponte Olivo airfield complex. Higher level planners also determined that the airborne assault would take place at night roughly two hours before the 0245 hours, 10 July, time set for H-Hour.

The Gela Plain, Ponte Olivo airfield, and the Piano Lupo (distance).
From page 186, *Sicily and the Surrender of Sicily.*

Within the parameters described above, division planners had essentially a free hand to decide how best to accomplish the assigned missions. The major problem that they confronted was a lack of lift (250 aircraft of the 52d Troop Carrier Wing, 64th Troop Carrier Group, and 316th Troop Carrier Group) which restricted the initial drop to four battalions plus attachments. MG Ridgeway tasked COL James Gavin's 505th PIR (reinforced) with accomplishing the division's initial objectives while the rest of the 504th stood ready for parachute insertion on D+1 or D+2 and the balance of the artillery and 325th Glider Infantry Regiment to arrive by glider (if available) in the third lift. The 505th task force totaled 3,405 men and comprised three battalions of the 505th Parachute

Infantry, Third Battalion of the 504th Parachute Infantry, three batteries of 456th Parachute Field Artillery Battalion with 75mm pack howitzers, Company B of the 307th Airborne Engineer Battalion, a detachment of the 82d Airborne Medical Company, air and naval gunfire support parties, and prisoner of war interrogation personnel. Division planners came up with a simple direct plan to seize the Piano Lupo. The regimental headquarters, First and Second Battalions of the 505th, and two batteries of the 456th would land north of a key highway intersection called "the Y junction," seize it, and prepare to meet any counterattacks that came their way. The 3/505th and C/456th would land south of the Y junction and seize the high ground overlooking it. Meanwhile, 3/504th would land south of the town of Niscemi and block the roads leading out of it. A small detachment equipped with demolitions would land near the Ponte Dirillo road and railroad crossings of the Dirillo (or Acate) River, demolish them, and set up a roadblock until infantry from 45th Division's 180th Regimental Combat Team relieved them. Although there was no time for a full scale rehearsal, Gavin, his battalion commanders, and the air group commanders flew the route under lighting conditions similar to those expected on the night of the drop and were able clearly to identify key checkpoints and terrain features. Planners, within the confines of operational security, also made clear to key leaders how the airborne plan fit into the intent of the larger army mission of establishing a solid foothold ashore. In a note distributed just prior to emplaning for Sicily, Gavin translated the essence of the mission in terms that any trooper could understand, "Let us carry the fight to the enemy ... Attack violently. Destroy him wherever found." The soul of the mission was, as Gavin recognized, engaging the enemy to help out the men who would soon be wading ashore.

As the first aircraft took off from airfields near Kairouan, Tunisia, at 2015 hours, 9 July, few could have imagined that the detailed drop plan was about to fall apart. The night flight to Sicily required several course changes, mandated radio silence, and required low-level flying to avoid possible radar detection. Completing the requirements of the troop carrying mission would have been difficult in the best of conditions and in a crosswind that blew from the northwest at 25 to 35 miles per hour, it proved a challenge that few of the combat inexperienced aircrews could master. The result was a drop that, with few notable exceptions, scattered paratroopers far and wide. At least four hundred landed behind the British beaches where some helped capture the towns of Avola and Noto, 65 miles from their intended drop zones. Many sticks were still recorded as missing over a month later. Most, however, found themselves far to the east of the Piano Lupo in the 45th Division's zone and on terrain that bore little resemblance to that which they had studied on maps and aerial photographs back in North Africa. Perhaps 12 percent of the force, Gavin later estimated, landed anywhere near according to plan. Fortunately, those few 505th and 504th troopers that did drop on or near their intended DZs were able to seize the Y junction and block the road south from Nescemi after severe fighting and thereby preventing a coordinated corps-level counterattack from developing against the Gela beachhead on D+1. Consequently, soldiers from 1st Division and rangers primarily faced counterattacks by the Italian 4th *Livorno* Division as they established themselves ashore rather than facing the combined might of both it and German *Hermann Goering* Division. In the darkness all over southeastern Sicily, individual troopers rolled up their sticks, formed small groups, and set about determining where they were prior to following Gavin's instructions to carry the fight to the enemy. Although the abortive drop made fully organized activity impossible, the actions of these widely-scattered extemporized guerrilla bands caused confusion among the enemy. For the 45th Division, the unintended screen of paratroops proved an unexpected boon and assisted the division in establishing itself ashore. For example, one large group from MAJ Mark Alexander's 2/505th captured Santa Croce Camerina, a 45th Division objective.

Paratroopers preparing to emplane for Sicily, 9 July 1943.

From page 116, *Sicily and the Surrender of Sicily.*

COL James Gavin's experience on D+1 and D+2 encapsulated that of many of his paratroopers. Gavin landed hard 30 miles southeast of the

DZ. He had no idea where he was, had no radio, suffered a leg injury upon landing, and soon found him among a small group of paratroopers. For the moment, the task force commander controlled 20 men. Gavin later wrote that gun flashes to his northeast at least assured him that he was indeed in Sicily. In doubt about exactly what to do but resolved to do something useful, he decided to move to the sound of the guns, a maxim remembered from his West Point days. Thus his tiny band set out northwest. Despite his injury, Gavin set a blistering pace so that when dawn arrived six hours later, only six men remained with him. The group then ran into a platoon strongpoint and took small arms and mortar fire, losing one man. Gavin's band returned fire, evaded, moved cross-country, and took cover. The COL spent one the longest days of his life hiding in a ditch with CPT Ben Vandervoort, the regimental S-3. Because of the danger of stumbling into another firefight while moving across exposed terrain, Gavin spent a sleepless, frustrating, and infuriating day ruminating over the fate of his command. Clearly the jump plan was in ruins but was the task force destroyed as a result? He considered his first day as a combat leader a failure, which affirmed a steely determination to find his command and engage the enemy. At nightfall, the group set out to the northwest where they bumped into a cluster of wounded and injured 505th troopers and, at 0230 hours five miles southwest of Vittoria, a 45th Division outpost. For the first time Gavin now knew exactly where he was, fifteen miles from the Piano Lupo. The group, now numbering eight, continued on to Vittoria where Gavin borrowed a jeep. He set out west along the highway to Gela and quickly came upon roughly 250 men, primarily from LTC Edward Krause's 3-505th that had fortified a tomato field. The day before, the regimental executive officer, LTC Herbert Batcheller, had ordered Krause to cease marching west and to dig in where he was. Gavin now angrily ordered Krause, who had failed to emplace outposts around his position, to get moving. Gavin sensed that something big was happening and that time was not on the paratroopers' side. Consequently, he drafted a platoon of airborne engineers and hurriedly set out west to reconnoiter. Soon the 100 or so foot tall Biazza Ridge (Sicilians call it Biazzo Ridge after the Villa Biazzo on its crest) loomed in the distance. After over 48 hours without sleep and an arduous cross-country odyssey, Jim Gavin's driving desire to rejoin his task force and to get into the fight placed him at a crossroads in his career as a combat leader.

    Gavin's instincts were excellent but the first indication that *Generalleutnant* Paul Conrath's *Hermann Goering* Division was on the move was inauspicious. Gavin's group captured a German officer on a motorcycle who told them that he had come down the road from Biscari. That could only mean that Highway 115, which pointed like a dagger into the flanks of both the 1st and 45th Division landing zones, was open at least as far as the intersection with the Biscari road. At about the same time, the sound of intense firing echoed up ahead.

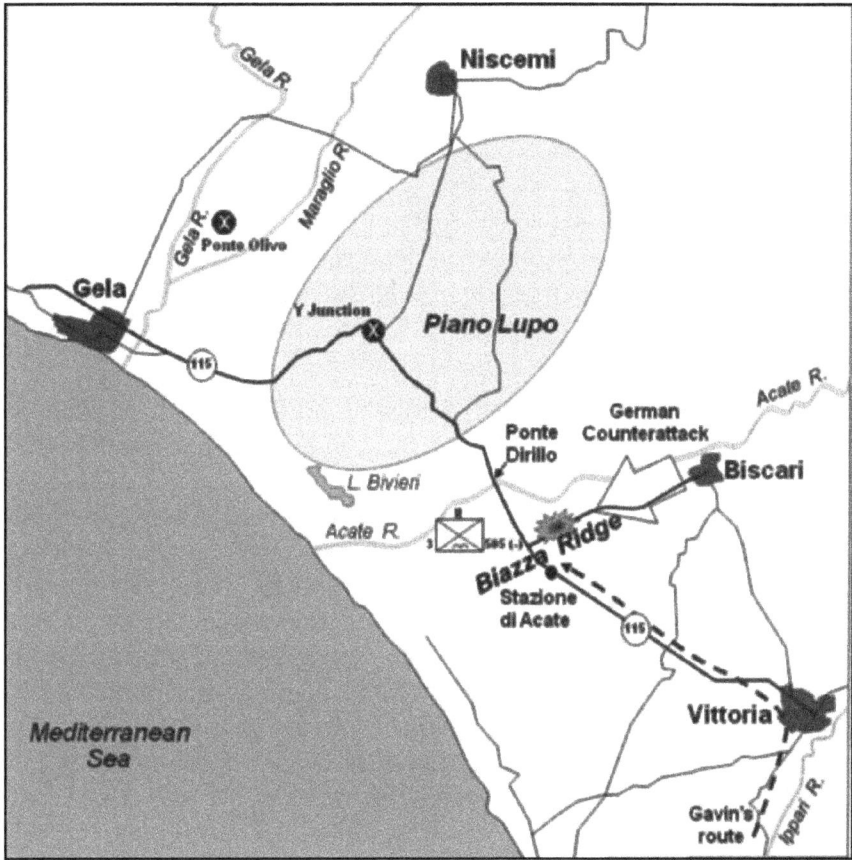

Figure 2. Movement and action on day 2 of operations in Sicily.

Gavin quickly pushed on about a mile and, at 0830 hours, arrived at *Statione di Acate*, a small train station where a railroad crossed Highway 115. Less than a mile ahead rested the gently sloping mass of Biazza Ridge. Gavin ordered the engineers to seize the ridge and led the way. As they approached the top of the ridge, firing became intense, and 1LT Benjamin Wechsler, the leader of the engineer platoon, was wounded. Gavin ordered the engineers to hold their ground and returned to the station where he met the XO of 3-505th who advised that the battalion would soon arrive and that LTC Krause had gone to request support from 45th Division. Soon about 220 paratroopers arrived with a platoon of the 3-180th and a couple members of a naval fire support party in tow. Gavin ordered a hasty attack which culminated as it crested the hill.

It began to dawn on Gavin that he was confronting a major German formation. Indeed, he was. The *kampfgruppe* assailing the *ad hoc* force of Americans was the eastern part of a three-pronged *Hermann Goering* Division attack on the Gela landing that the paratroops on the Piano Lupo had helped foil on D+1. This same task group, composed of a company of heavy Mark VI (Tiger) tanks, a few medium Mark IV tanks, two battalions of panzer grenadiers, and an armored artillery battalion, had attempted to force the Ponte Dirillo crossing on D+1 but were repulsed by Company G, 505th and elements of the 180th Infantry. Significantly, the powerful *kampfgruppe* returned up the road to Biscari to regroup only after savaging the 1-180th, capturing its battalion commander and over two hundred prisoners. Now with a new and more aggressive commander, the German unit returned to the site of its earlier defeat with renewed determination and ferocity.

With an *ad hoc* command composed primarily of light infantry, COL Gavin assessed the risk to his force in terms of the overall intent of his mission of engaging enemy reserves before they had a chance to attack the beachhead. Clearly, the Germans had the advantage in terms of equipment, firepower, and numbers. A decision to retreat toward Vittoria to consolidate with better-equipped elements of the 45th Division could be justified as there was certainly no guarantee that the Germans would not do serious damage to Gavin's force as they had the 1-180th the day before. However, Gavin determined that he could best contribute to the 7th Army's fight here and ordered his men to dig in and stand firm. The after-action report of the 505th makes clear Gavin's thought process, "It appeared evident that the Ridge dominated the area between the ACATE River and VITTORIA and its loss would seriously jeopardize the landings of the 45th Division. It was decided to hold the Ridge at all cost and if the tanks entered the defense, to destroy the infantry accompanying them." Left unsaid was the fact that troops on Biazza Ridge controlled access to the undestroyed bridge at Ponte Dirillo on Highway 115. Thus, possession of the ridge also protected the right flank of hard-pressed 82d and 1st Division units battling the Germans on the Piano Lupo on D+2.

Gavin's decision resulted in some of the bitterest fighting of the Sicilian campaign. The fighting ebbed and flowed all day as the German's repeatedly attacked the American position. Gavin's troops struggled to scratch shallow foxholes in the hard-baked rocky ground. The troops possessed few heavy weapons. When direct fire proved ineffective against heavily armored Tiger tanks, Gavin's men improvised by firing at the soft underbellies of the behemoths as they climbed over rises. On more than one occasion, German tanks drove among the Americans' fighting positions, engaging individual soldiers with their main guns but the men stood firm and stripped away the armor's infantry support which caused each attack to culminate. Gavin's blood was up and his men drew inspiration from his determined hands-on leadership. Indeed, he later wrote that he saw the battle as an opportunity to test his paratroopers against "the toughest opposition we could find." They would not be found wanting.

The battlefield in 2012: the Acate (Dirillo) River valley, Ponte Dirillo, and Biazza Ridge (distance) as viewed from the northwest.
Author's collection.

Gavin and his men's grit and determination made a victory possible, but left unsupported, attrition would gradually have taken its toll. Fortunately, the tiny force did not have to carry the day alone and American strength gradually grew as the day wore on. Other paratroopers marched to the sound of the fighting. For example, two 75mm pack howitzers of LTC Harrison Harden's 456th Parachute Field Artillery Battalion arrived and were immediately pressed into service, occasionally dueling over open sights with German tanks. Harden himself had dropped 32 miles from the assigned DZ. Later, two 57mm anti-tank guns from the 180th Infantry added their weight to the fight. During the afternoon, the attached naval gunfire support party was able to make contact with cruisers and destroyers offshore. Their 5- and 6-inch salvos served to break up more than one attack. In total, naval vessels fired over 1,800 rounds of high explosive in support of Gavin's force. So, too, did 45th Division 155mm howitzers manifestly contribute to the defense. At 1800 hours, more paratroopers arrived accompanied by a company of 45th Division Sherman tanks. This proved the turning point of the battle as Gavin now decided to counterattack. In the battle's final major action, the strengthened force, which included Gavin later wrote, "regimental cooks, clerks, truck drivers, everyone who could carry a gun," drove the Germans off the ridge and back in the direction of Biscari while capturing several trucks, many heavy weapons, a few tanks, and 12 120mm mortars in the process. Because of COL Gavin's leadership and the matchless efforts of his men, Biazza Ridge was by nightfall on D+2, firmly in American hands.

COL James Gavin with war correspondent "Beaver" Thompson, who jumped with the 82d, at *Statione di Acate* near Biazza Ridge, 11 July 1943.

From page 169, *Sicily and the Surrender of Sicily*.

A day had made all the difference. Exhausted after 60 sleepless hours of almost continuous exertion, Jim Gavin could finally relax a bit and reflect that he and the paratroopers and soldiers under his command had indeed passed the test of combat.

For Further Reading

Rick Atkinson. *The Day of Battle: The War in Sicily and Italy, 1943-1944*. New York: Henry Holt, 2007.

William B. Breuer. *Drop Zone Sicily: Allied Airborne Strike, July 1943*. Novato, CA: Presidio Press, 1983.

Carlo D'Este. *Bitter Victory: The Battle for Sicily*. 1988; paperback edition, New York: HarperCollins Publishers, 1991.

James M Gavin. *On to Berlin: Battles of an Airborne Commander, 1943-1945*. New York: Viking Press, 1978.

Ed Ruggero. *Combat Jump: The Young Men who Led the Assault into Fortress Europe, July 1943*. New York: HarperCollins Publishers, 2003.

John C Warren. *Airborne Missions in the Mediterranean, 1942-1945*. USAF Historical studies: No. 74. Maxwell, AL: USAF Historical Division, Research Studies Institute, Air University, 1955.

The Six Principles of Mission Command

1. Build Cohesive Teams through Mutual Trust

2. Create Shared Understanding

3. Provide a Clear Commander's Intent

4. Exercise Disciplined Initiative

5. Use Mission Orders

6. Accept Prudent Risk

Mission Command Illustrated in the Biazza Ridge case

1. Build Cohesive Teams through Mutual Trust. Once in North Africa, personnel reassignments, except for those due to training injuries, ceased. This stabilized unit rolls and, when combined with a rigorous and ambitious training program, enabled leaders to build cohesive teams and mutual trust. The shared hardship of living together in the harsh North African climate under primitive conditions also served to build unit cohesion.

2. Create Shared Understanding. Unit commanders clearly understood the purpose of the airborne drop within the invasion plan: to facilitate the establishment of a firm lodgment ashore by engaging Axis reserves moving toward the beachhead. COL Gavin translated the overall intent to his men in plain language in a note issued just prior to embarkation. Although all units understood their role in the detailed plan, when the drop went awry, "little groups of paratroopers" understood that engaging the enemy wherever found was the soul of the mission.

3. Provide a Clear Commander's Intent. The division's plan made clear the intent of the drop and the missions for each element. 82nd Airborne planners had essentially a free hand to decide how best to meet the intent handed down from higher headquarters; the division was assigned a sector for the drop rather than specific objectives. Furthermore, Gavin's pre-drop message drove home his intent to the troops—"Let us carry the fight to the enemy ... Attack violently. Destroy him wherever found." This was something that any paratrooper, wherever he landed, could understand. Gavin's efforts in this regard paid dividends when the drop plan fell apart.

4. Exercise Disciplined Initiative. Airborne training in North Africa made clear that doing something proactive when in an ambiguous situation was expected and required. Individual and unit training provided

troopers with the ability and incentive to operate effectively in *ad hoc* organizations. Training in things such as small unit tactics, cutting communication lines, night navigation, and with enemy weapons also provided troopers with the skills to succeed as well as the confidence to take action. All over southeastern Sicily, paratroopers seized opportunities to make a contribution to the success of the amphibious assault rather than waiting for orders. COL Gavin clearly took it upon himself to make things happen. His efforts to get both himself and dispersed 82nd combat power to the Piano Lupo, the division objective, led directly to the battle at Biazza Ridge, which greatly contributed to the success of the landings.

5. Use Mission Orders. The 82nd Airborne Division issued formal orders prior to the drop with detailed schemes of maneuver. This being said, levels of command higher than division did not prescribe how to accomplish the division's mission and allowed subordinate commands latitude to decide for themselves. Once in Sicily, orders were usually verbal and communicated intent rather than prescribing specific action. Gavin's orders to LTC Krause to march west on D+2 and to the engineers on Biazza Ridge to hold firm were examples of this. In neither case, did the task force commander delineate exactly how his intent was to be accomplished.

6. Accept Prudent Risk. The very real threat of counterattack in the Gela sector justified the airborne drop. Only the use of airborne troops promised to prevent or at least disrupt a counterattack against the most vulnerable portion of 7th Army's landing zone at the most critical time—initial debarkation. At Biazza Ridge, Gavin concluded that the risk to the landing zone posed by the *kampfgruppe* of the *Hermann Goering* Division justified the risk to his tiny force. The bold decision to hold at Biazza Ridge clearly met the intent of the parachute drop in the first place although the plan did not anticipate major action there.

Thunder Run in Baghdad, 2003

Anthony E. Carlson, Ph.D.

MG Buford "Buff" Blount faced a critical decision. During the previous two weeks, his 3d Infantry Division (ID) (Mechanized) had raced 700 kilometers through southern Iraq, reaching the outskirts of Baghdad in early April 2003. The division had overrun both Baghdad's airport west of the city (Objective LIONS) and the key intersection of Highways 8 and 1 (Objective SAINTS) directly south of the city, allowing it to create a partial cordon around the capital. Blount and the senior leaders of US Army V Corps, 3d ID's higher headquarters, now needed to seize the city and collapse Saddam Hussein's regime, but how?

Blount and V Corps Commander LTG William S. Wallace had no concrete intelligence about the capability and intent of the Iraqi forces protecting Baghdad. To collect intelligence about the conventional and paramilitary units inside the city, they planned an armored reconnaissance in force. At 1600 on 4 April, Blount gave the mission to COL David G. Perkins, commander of 3d ID's 2d Brigade, for execution the following morning. Staging out of Objective SAINTS, the battalion-sized column of M1A1 Abrams tanks and M2 Bradley Fighting Vehicles would attack north on Highway 8 into the middle of western Baghdad and then turn west, linking up with COL William Grimsley's 1st Brigade, 3d ID, at the airport. The bold plan, which Wallace judged a "reasonable risk," was destined to become the first armored foray into a major city since World War II.

Perkins assigned the so-called "thunder run" mission to LTC Eric Schwartz's Task Force (TF) 1st Battalion, 64th Armor Regiment (1-64 AR). Schwartz's TF 1-64 AR included 731 Soldiers, 30 M1A1 tanks, 14 Bradley infantry fighting vehicles, 14 engineer vehicles, and other mechanized support vehicles. Perkins' intent was to attack up Highway 8 to "create as much confusion as I can inside the city because I had found that my Soldiers or my units can react to chaos much better than the enemy can." Although the sudden new mission caught Schwartz off guard, he praised the straightforward commander's intent and purpose. "The planning was simple," he explained. "The thunder run mission was the simplest of all tasks that we were given. There was no maneuver required. It was simply battle orders followed by battle drills."

At 0600 on 5 April, Schwartz's armored column rolled north up Highway 8. In the vanguard of the staggered column was CPT Andrew

Hilmes' Alpha Company. COL Perkins accompanied the task force in his command M113 armored personnel carrier to observe firsthand the effectiveness and distribution of enemy forces.

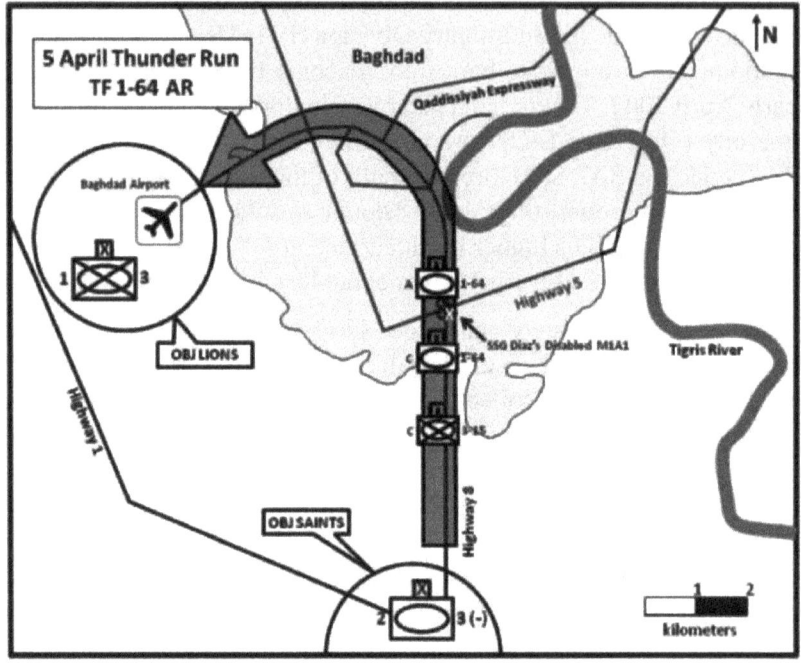

Figure 1. 5 April Thunder Run, TF 1-64 AR.

Moments after beginning the movement, the task force came under intense and sustained fire. Special Republican Guard (SRG) soldiers, *Fedayeen Saddam* militiamen, Syrian and Palestinian mercenaries, and other paramilitary forces unleashed an unremitting barrage of AK-47 rifle fire, rocket propelled grenades (RPGs), and mortar rounds from hastily-prepared positions adjacent to the highway. As the task force rumbled north, police cars, taxis, ambulances, garbage trucks, and other civilian vehicles massed along the highway, depositing hundreds of additional enemy fighters. The rifle and RPG volleys turned the operation into something akin to running a gauntlet of fire but it did little to slow the armored column.

TF 1-64 AR Attacking up Highway 8 on 5 April 2003.
Photo Courtesy of the Fort Stewart Museum, US Army.

Near the first overpass on Highway 8, an RPG round exploded in the rear of SSG Jason Diaz's tank, immobilizing it. As Diaz's crew struggled to put out a growing fire and get the disabled tank rolling again, trailing Abrams and Bradley Fighting Vehicles formed a defensive perimeter. The tankers mowed down dozens of fighters assembling alongside the highway with coaxial machine gun fire and main gun rounds. Since Perkins' order emphasized momentum, LTC Schwartz made the call after half an hour to abandon Diaz's tank, recover the crew, retrieve sensitive computer systems, and attack north deeper into the city.

The armored column passed the Qaddissiyah Expressway ramp towards downtown Baghdad and turned west in the direction of the airport, entering crowded residential neighborhoods. Hundreds of paramilitary fighters and military personnel assaulted Schwartz's column from all directions, only to fall victim to the Americans' overwhelming firepower. The enemy resorted to placing makeshift concrete barriers across the highway and even launching suicide vehicle attacks but with no success. After two hours and 20 minutes, the column arrived at the airport. COL Perkins concluded that the reconnaissance in force had completely surprised the regime. "[The Iraqis] thought that they could bloody our nose enough on the outside of the city … that we just would not push through block by block," Perkins explained. "They weren't planning for this very heavy armored thrust busting right through, coming in[to] the city."

The thunder run demonstrated that US armored forces could penetrate Baghdad while suffering minimal casualties. During the movement, TF 1-64 AR sustained one destroyed Abrams tank, one heavily damaged Bradley, one Soldier killed in action (KIA), and four Soldiers wounded in action (WIA). Schwartz's task force killed at least 1,000 Iraqi and Syrian fighters, destroyed 30 to 40 Russian-manufactured BMP infantry fighting vehicles and other vehicles, destroyed one T-72 main battle tank, and eliminated countless roadside bunkers. The operation provided excellent indicators of enemy tactics, strength, and fighting positions. For instance, the task force discovered that the enemy preferred to mass fires from overpasses. Perkins observed that the bridges provided the enemy cover and concealment and afforded "avenues of approach in the flank."

LTG Wallace and MG Blount praised the 5 April thunder run. They envisioned it as a prelude to additional armored missions in and out of the city that would disrupt Baghdad's defenses with the paramount goal of regime collapse. Late on 5 April, Wallace ordered a second such mission for 7 April. Blount again assigned the task to 2d Brigade.

After returning to SAINTS with TF 1-64 AR and receiving Blount's orders, Perkins proposed a bolder course of action to his division commander. He wanted to take two armor task forces into Baghdad and *turn east* at the same intersection where TF 1-64 AR had looped west towards the airport. The task forces would travel several additional kilometers and occupy the regime's downtown government complex on the banks of the Tigris River, the location of Saddam Hussein's ornate palaces, his ruling party's headquarters, parade grounds, and war monuments. With the rest of V Corps and the 1st Marine Expeditionary Force bearing down on Baghdad from southwest and southeast respectively, Perkins identified the downtown palaces as the regime's "center of gravity." He hoped to avoid an endless cycle of armored forays that scored tactical victories but did not hasten strategic success.

Perkins also feared that the US Army was losing the information war. The Iraqi information minister, Mohammed Saeed al-Sahhaf, had taken to the airwaves and falsely announced that Iraqis had slaughtered US Soldiers outside of Baghdad. To make matters worse, the British Broadcasting

Company was broadcasting al-Sahhaf's propaganda to the world. Perkins wanted to send an unmistakable message to Iraqis that the regime's days were numbered. "I didn't want [the false stories] to happen again," he emphasized. "[Al-Sahhaf's disinformation was] falsely emboldening the Iraqis to continue to fight and defend [the city] ... stretching this war out." Perkins concluded that the enemy's relatively unsophisticated and uncoordinated resistance during the first thunder run showed that such a bold operation was possible.

On 6 April, Blount brought Perkins' recommendation before LTG Wallace. The corps commander dismissed it. Even though Wallace sought to render the regime "irrelevant," the plan at Combined Forces Land Component Command (CFLCC) level at this point intended to topple the regime through synchronized attrition rather than a dramatic armored thrust. The CFLCC envisioned creating a cordon of forward operating bases (FOBs) around Baghdad from which US forces could launch pinpoint raids and seize critical objectives so that they did not have to clear the city block by block. From a tactical perspective, Wallace also feared that Perkins might overextend his line of communication (LOC) between Objective SAINTS and the palace grounds, isolating the task forces in a hostile city of five million people without the ability to resupply his units or evacuate casualties. He directed Blount to take a "less aggressive tactic" that involved attacking into the city to the point of the airport interchange but then returning to SAINTS.

The events that unfolded over the next 24 hours serve as a clear illustration of mission command principles in action. As Perkins prepared to execute V Corps' limited objective for the second thunder run, he conceptualized an additional plan to allow 2d Brigade and its assigned units to go downtown and "stay the night" if conditions warranted. Privately, Perkins set four preconditions to meet before he would offer his option to go downtown and stay during the mission. The preconditions were based on "lessons learned" during the first thunder run:

1. The 2d Brigade could successfully fight its way into downtown without becoming fixed.

2. Seizing defensible and symbolic terrain at the downtown palace complex.

3. Opening and maintaining a ground LOC using Highway 8 and the Qaddissiyah Expressway between the Tigris River and Objective SAINTS.

4. Logistical conditions supported remaining overnight.

On the afternoon of 6 April, Perkins briefed his intent. Speaking in a dusty tent without notes, slides, or handouts, Perkins explained to his subordinate commanders that the entire brigade would conduct a second thunder run at dawn the next morning. He instructed them to prepare to spend the night downtown. "We have set the conditions to create the collapse of the Iraqi regime. Now we're transitioning from a tactical battle [*sic*] to a psychological and informational battle," he said. Maintaining momentum during the movement was paramount. "Attack as fast as you can, and push right through to the center of the city," Perkins added. "If a vehicle becomes disabled due to enemy fire, you immediately take the crew off, put them on another vehicle, and you just leave it."

The scheme of maneuver had LTC Schwartz's TF 1-64 AR assuming the vanguard. If conditions warranted turning northeast towards downtown, TF 1-64 AR would seize downtown Objective DIANE, which included the Tomb of the Unknowns, a park, and a zoo. LTC Philip Draper deCamp's TF 4th Battalion, 64th Armor Regiment (TF 4-64 AR), would follow TF 1-64 AR and seize two of Saddam Hussein's palaces on the Tigris River (Objectives WOODY EAST and WOODY WEST). The third battalion, LTC Stephen Twitty's TF 3d Battalion, 15th Infantry Regiment (TF 3-15 IN), would keep the LOC open between Objective SAINTS and downtown. To do so, TF 3-15 IN had to control three vital overpass intersections on Highway 8, designated as CURLY, LARRY, and MOE. MOE was the key interchange where Perkins' Soldiers either had to move east in the direction of downtown or make a U-turn, returning to SAINTS. For Perkins, controlling the three overpass intersections was decisive to securing MG Blount's approval of his option to go downtown.

The second thunder run got off to a rocky start. In the wake of the 5 April attack up Highway 8, the Iraqis had laid a minefield on the highway north of SAINTS, extending for 500 meters. At 0538 on 7 April, CPT David Hibner's company of 2d Brigade engineers hastily cleared 444 mines. By 0600, TF 1-64 AR, TF 4-64 AR, and TF 3-15 IN departed in that order in a long column. Only eleven minutes into the movement, enemy small arms fire, RPGs, and mortar rounds erupted from both sides of the highway. In accordance with COL Perkins' intent, the two leading task forces continued to advance and hand over targets to trailing units, which also recovered the crews of disabled armored vehicles.

Perkins faced his first critical decision an hour into the operation. As the armored column clanked towards MOE, he radioed BG Lloyd J. Austin III, Assistant Division Commander (Maneuver), explaining that the level of resistance faced by 2d Brigade was less intense than during the previous thunder run. He stated his preconditions for going downtown, insisting that he could meet all of them. Without giving a definitive answer, Austin stated that he would inform Blount. He told Perkins to continue the advance and see how the fight developed. Shortly after 0700, the armored column turned east off Highway 8 and, within an hour, seized DIANE, WOODY EAST, and WOODY WEST. The brigade commander calculated that he had enough fuel to delay a final decision about formally requesting an overnight stay until 1000. In his mind, the shock value of keeping US armor task forces downtown outweighed the significant risks associated with being isolated in a hostile urban environment.

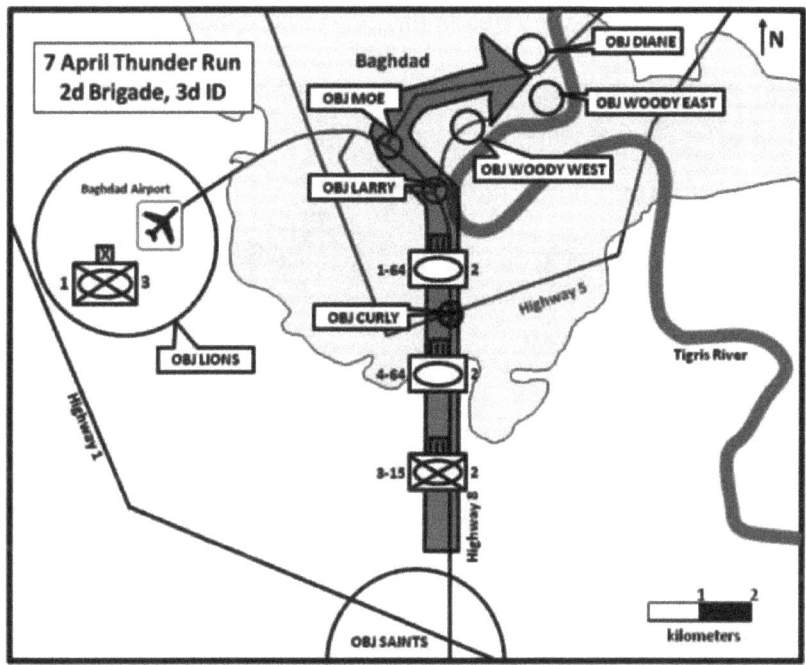

Figure 2. 7 April Thunder Run, 2d Brigade.

The movement off Highway 8 caused a stir at V Corps headquarters. When LTG Wallace went to bed on 6 April, he thought that 2d Brigade would advance to MOE and then make a U-turn, heading back to SAINTS. As the armored task forces advanced towards the downtown objectives, Wallace observed the operation on the screen of his Blue Force Tracker. Stunned, the corps commander asked Blount about the unexpected deviation from his intent during their regular morning brief. Blount explained Perkins' estimation that the diminished resistance justified turning downtown and positioning tanks at Hussein's palace complex in a dramatic show of the regime's irrelevance. Tension filled the room as Wallace contemplated the situation. Finally, Wallace broke the long silence by signaling his eager approval. According to COL Russell Thaden, the V Corps Deputy G2 (intelligence officer) who was present at the time of the conversation, Wallace replied, "Go ahead, I trust your judgment. If you think you can get to the palace and hold it, [its] your call and I'll clear it [with CLFCC.]" Refusing to focus on the divergence from his original guidance, Wallace instead recognized that one of his subordinate commanders had created an opportunity for success through disciplined initiative and prudent risk taking. He believed that the overall result of the mission was more important than the methods used to achieve it. Both the corps and division commanders therefore deferred to the judgment of the commander on the ground.

Attacking towards Downtown Baghdad on 7 April 2003.
Photo Courtesy of Fort Stewart Museum, US Army.

Meanwhile, the 2d Brigade faced a rapidly deteriorating situation. As TF 3-15 IN slugged it out at CURLY, LARRY, and MOE with bands of determined enemy fighters, a rocket attack disrupted the brigade tactical operations center (TOC) at Objective SAINTS, killing three Soldiers and temporarily cutting off communications. In the midst of the mayhem, LTC Eric Wesley, the 2d Brigade executive officer (XO), calmly orchestrated efforts to triage wounded Soldiers and evacuate disabled vehicles. Within 45 minutes, Wesley had reestablished communication and set up a makeshift TOC, minimizing the disruption of command and control. Perkins praised Wesley and all Soldiers at the TOC for remaining focused on the mission in the midst of disarray. He later expounded on the Soldiers' shared understanding of his intent, "Everyone understood how important it was to stay in the city and not have to fight the fight again."

Events continued to spiral out of control on Highway 8. As Perkins reached his self-imposed 1000 deadline for seeking permission to spend the night, TF 3-15 IN was still fighting to maintain control of the three interchanges at CURLY, LARRY, AND MOE. Even worse, Iraqi fighters ambushed the first convoy of heavy expanded mobility tactical trucks (HEMTTs) hauling much-needed supplies and fuel up Highway 8. Five HEMTTs were destroyed, two Soldiers killed, and Highway 8 remained disputed. Fierce fighting around Objective MOE also left a mechanized infantry company critically short of ammunition.

MG Blount, LTG Wallace, and COL Perkins in Baghdad, April 2003.
Photo Courtesy of Fort Stewart Museum, US Army.

Despite the dire circumstances, COL Perkins refused to rush his decision. "If you had a decision matrix," he stated, "it probably d[id] not pay to spend the night." Nevertheless, he delayed because he did not want to surrender symbolic ground or face the possibility of ordering additional armored attacks in the coming days. Withdrawing from the city would also embolden the regime and provide additional propaganda for the information minister. Based on extensive pre-war training in Kuwait, Perkins trusted LTC Twitty's task force to win the battles at the overpass intersections if given sufficient time, bought by delaying a decision past 1000. To mitigate resupply problems, he instituted an "energy conservation plan," ordering TF 1-64 AR and TF 4-64 AR tank commanders to turn off their engines. He then positioned the task forces' Bradleys at key downtown bridges and intersections to strengthen the defensive posture. Perkins believed that such measures would buy him several additional hours before supply concerns might force him to withdraw.

MG Blount again trusted the judgment of his commander on the ground. At 1016, he reinforced TF 3-15 IN by moving the 1st Brigade's TF 2d Battalion, 7th Infantry Regiment (TF 2-7 IN), to occupy and defend Objective CURLY, allowing TF 3-15 IN to focus on clearing LARRY and MOE. By late afternoon, the infantry task forces had defeated the Iraqi fighters along Highway 8 and cleared the LOC for the HEMTTs to move north to supply Perkins' brigade.

Just hours before sundown, the fuel and ammunition resupply reached downtown after a harrowing movement up Highway 8. COL Perkins' deliberate decision-making and confidence in his subordinate commanders validated LTG Wallace's and MG Blount's trust in him. By early evening, Wallace approved the decision to spend the night.

There is always a tension between executing mission orders and exercising disciplined initiative but Wallace clearly understood the benefits of empowering subordinate commanders to make decisions in a fluid, complex, and highly unpredictable tactical environment. "COL Perkins, to his credit … was taking advantage of the situation that was presented to him on the battlefield," Wallace explained, "which is what we teach our young leaders to do." Ultimately, the second thunder run produced tactical, strategic, and information victories as television networks soon broadcasted images of US tanks occupying Saddam Hussein's former seat of power. In retrospect, Perkins attributed the 2d Brigade's success to the flexibility displayed by Wallace and Blount and their willingness to empower him with freedom of action:

> These thunder runs were successful because the corps and division-level commanders established clear intent in their orders and trusted their subordinates' judgment and abilities to exercise

disciplined initiative in response to a fluid, complex problem, underwriting the risks that they took.

The Iraqi information minister could no longer deny that US Soldiers occupied Saddam Hussein's seat of administrative power. The regime teetered on the brink of an inevitable collapse. Within weeks, the Baathist government no longer ruled Iraq.

For Further Reading

CPT Jason Conroy. *Heavy Metal: A Tank Company's Battle to Baghdad*. Dulles, VA: Potomac Books, 2005.

COL (retired) Gregory Fontenot, LTC E. J. Degen, and LTC David Tohn. *On Point: The United States Army in Operation Iraqi Freedom Through 01 May 2003*. Washington: Office of the Chief of Staff, US Army, 2004.

Jim Lacey. *Takedown: The 3rd Infantry Division's Twenty-One Day Assault on Baghdad*. Annapolis: Naval Institute Press, 2007.

LTG David G. Perkins. "Mission Command: Reflections from the Combined Arms Center Commander." *Army* 62 (June 2012): 30-34.

COL (retired) John B. Tisserand III. *US V Corps and 3rd Infantry Division (Mechanized) during Operation Iraqi Freedom Combat Operations (March to April 2003)*. Vol. 3 of *Network Centric Warfare Case Study*. Carlisle Barracks, PA: Center for Strategic Leadership, 2006.

David Zucchino. *Thunder Run: The Armored Strike to Capture Baghdad*. New York: Grove Press, 2004.

The Six Principles of Mission Command

1. Build Cohesive Teams through Mutual Trust

2. Create Shared Understanding

3. Provide a Clear Commander's Intent

4. Exercise Disciplined Initiative

5. Use Mission Orders

6. Accept Prudent Risk

Mission Command in the Thunder Run case

1. Build Cohesive Teams through Mutual Trust. After assuming command of 2d Brigade in June 2001, COL Perkins benefitted from a low turnover of battalion commanders, fostering stability, continuity, and mutual trust. A six-month period of intensive training in Kuwait prior to the invasion of Iraq in March 2003 also reinforced mutual trust and unit cohesiveness. Perkins trusted the collective capability of his Soldiers because he had seen them repeatedly participate in focused training missions. Later, he would describe mutual trust as "the bedrock of mission command."

2. Create Shared Understanding. The corps, division, and brigade commanders clearly conveyed their intents, objectives, and key tasks to subordinate commanders. For instance, during the thunder run missions, COL Perkins' battalion commanders understood that maintaining a high tempo and handing off targets to trailing armored vehicles and units was critical for mission success.

3. Provide a Clear Commander's Intent. Both LTG Wallace and MG Blount provided clear and concise commanders' intents for both thunder run missions. Their intent was to attack into Baghdad in an armored column to test Saddam Hussein's urban defense, collect intelligence about the paramilitary and military units, and maintain pressure on the regime. COL Perkins' intent closely mirrored those of his senior commanders. For the first thunder run, his intent was to attack up Highway 8 to "create as much confusion as I can inside the city, because I had found that my Soldiers or my units can react to chaos much better than the enemy can." Perkins' intent for the second thunder run was also clear and succinct. "You get on that road and you attack as fast as you can, and push right through to the center of the city," he stated. "If a vehicle becomes disabled due to enemy fire, you immediately take the crew off, put them on another vehicle, and you just leave it."

4. Exercise Disciplined Initiative. Perkins exercised disciplined initiative by creating an option for his task forces to go and stay downtown during the 7 April thunder run. As the armored column approached Objective MOE, Perkins assessed the situation, considered his options, and made a determination to go downtown. He believed that positioning two US armor task forces at the regime's seat of political and military power would expedite the accomplishment of Wallace's intent, which he judged more important than the specifics of the tactical operation. "The center of gravity for the regime is Baghdad," he stated. "If you get in there in what we call the regime district ... and if you could stay there, then no one could say that you're not there or that you're not in control over the city ... Saddam can still be alive, but he's irrelevant."

5. Use Mission Orders. Perkins issued written and verbal mission orders for both thunder run missions. For the first reconnaissance in force, he directed TF 1-64 AR to attack up Highway 8 all the way to the Baghdad Airport in order to collect intelligence about the composition and disposition of the Iraqi forces defending the city. The orders emphasized maintaining momentum, handing over targets to trailing armored vehicles, and avoiding pitched battles, but they also maximized individual initiative. Indeed, during the 5 April thunder run, Perkins did not micromanage the details of the movement. His mission order for the second thunder run included similar directions and guidance, with the two armor task forces attacking all the way downtown while TF 3-15 IN seized the three overpass objectives on Highway 8, opening the LOC between the TOC and Saddam Hussein's palace complex. As commander of TF 4-64 AR, LTC deCamp explained, "the mission was to bypass and not get into a pitched battle."

6. Accept Prudent Risk. Perkins accepted the risk of attacking into downtown Baghdad and spending the night because of two mitigating factors: his firsthand knowledge of the Iraqi resistance and the fulfillment of the four preconditions he had set for spending the night. Because he accompanied TF 1-64 AR on the 5 April mission, Perkins was able to conclude on 7 April that the Iraqi resistance had diminished in sophistication and coordination, justifying a turn downtown. Upon arriving at the downtown objectives, he mitigated the risk to his Soldiers by basing his decision to remain overnight on meeting the four preconditions. A V Corps after action review (AAR) briefing dated July 2003 praised the strategic implications of Perkins' prudent risk taking: "The decision to leave an armored brigade in the center of Baghdad overnight seemed unthinkable one day and obvious the next. We must never underestimate the psychological impact of an American armored force holding the ground it takes."

The Drive to Bastogne
Kendall D. Gott

During the Battle of the Bulge, the beleaguered 101st Airborne Division and Combat Command B of the 10th Armored Division were pressed into a tight perimeter as the German offensive swept around the key city of Bastogne. The American paratroopers were holding their own but supplies and ammunition could only come by airdrop and the foul winter weather hindered all efforts to deliver the needed materiel. Relief for the "battling bastards" would have to come from the ground. To do just that, the Third Army under LTG George S. Patton Jr. had shifted its attack to the east and sent its mobile divisions to the north. It was far from certain that Patton's armor and infantry divisions could reach Bastogne before the 101st Airborne was forced to surrender. The relief of Bastogne is a classic example of commanders at all levels using initiative and daring to overcome a determined enemy.

The German winter offensive through the Ardennes region in 1944 was designed to split the British and American Allied line in half, to capture the port of Antwerp, and then proceed to encircle and destroy four Allied armies, forcing the Western Allies to negotiate a peace treaty. Once that was accomplished, they could then fully concentrate on the eastern front against the armies of the Soviet Union. The offensive was planned in secrecy and surprise was achieved by combination of Allied overconfidence, preoccupation with their own offensive plans, and poor aerial reconnaissance. Although the ability of the Germans to assemble and organize the forces for this offensive was remarkable, these units were still beset with shortages in equipment, manpower, and logistics. Fuel was in critically short supply and the Germans were counting on rapid speed and capturing Allied fuel stocks to keep the momentum of the attack.

The secrecy of the German preparations had lulled Allied planners into believing this was a quiet sector of the front and had assigned relatively green units here to adjust them to combat or placed units in need of refitting or rest. The heavy overcast weather had grounded the far superior Allied air forces, preventing reconnaissance flights prior to the offensive and ground support once it commenced. When the German offensive began on 16 December, it stunned the Allied units in their path. Some were pushed back in disorder and others were simply overwhelmed by the onslaught. The Germans initially advanced quickly but stubborn resistance formed on the northern shoulder of the offensive around Elsenborn Ridge at the town of Hofen. In the south, the defenders in Bastogne blocked German access to key roads they were relying on for success. The stubborn defense by American units dug into the wooded hills covering the few good roads threw the German timetable behind schedule.

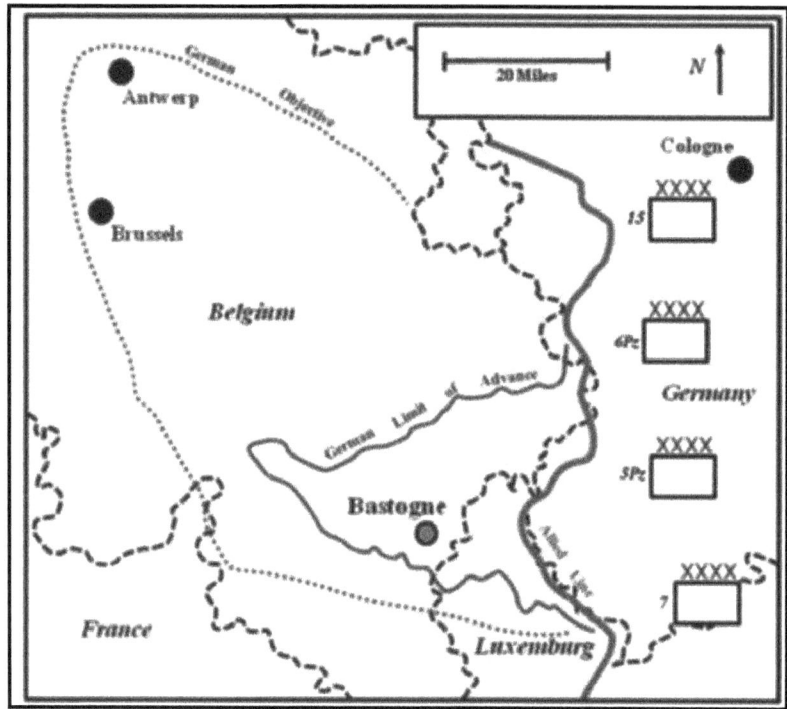

Map 1. The German offensive, December 1944.

The Germans were particularly anxious to seize the vital crossroads at Bastogne, where the American 101st Airborne Division was hurriedly deployed to block their advance. By 21 December they had contained the paratroopers in the town and surrounding hills while the *panzer* spearhead continued its drive to the west but as long as the Americans held Bastogne the Germans would have a very difficult time in resupplying their units and the men used to contain American perimeter were needed elsewhere. It became critical to the Germans to take Bastogne to keep the offensive going. It was critical to the Allies to hold Bastogne to disrupt the German plans.

The 4th Armored Division was part of this northern drive ordered by General Patton and, its commander, Major General Hugh Gaffey, was frustrated there was no quick breakthrough possible on the main Arlon-Bastogne highway. The Germans had expected such a thrust and had deployed strong defenses on this axis which slowed the advance to a crawl.

The heavy firepower of artillery and tanks were blasting the way clear for the infantry but it was still slow going. The common purpose of the division was to reach Bastogne as quickly as possible before the Germans could force its surrender.

The 4th Armored Division was task organized into three "combat commands" which were composed of two to four battalions and task-organized for the mission at hand. For the drive to Bastogne, Combat Commands A and B (CCA and CCB respectively) were the main efforts, while Combat Command Reserve (CCR), under the command of COL Wendell Blanchard, was given the task to screen the left flank of the division along the axis of advance. For this task, the nucleus of the CCR was the 37d Tank Battalion under LTC Creighton Abrams, and the 53d Armored Infantry Battalion under LTC George Jaques (pronounced Jakes). These two battalions were task-organized for this mission by swapping companies so that they were almost a 50-50 mix of tanks and infantry. Although this was not a permanent organization, the tank and rifle companies of CCR had teamed together many times in the past. Supporting units included the 94th Armored Field Artillery Battalion and a battery of 155mm howitzers from the 177th Field Artillery Battalion. Various other detachments of engineers, transport, and support units rounded out the organization.

Creighton Williams Abrams Jr. had graduated from West Point in 1936 and served with the 1st Cavalry Division in the years up to 1940. As the Armor branch formed, he transferred to that branch and served as a tank company commander in the newly formed 1st Armored Division. Abrams was reassigned to the 4th Armored Division and served as the regimental adjutant until June 1942. The 37th Tank Battalion was created during a reorganization of the division and Abrams was placed in command. During his command, the 37th Tank Battalion was used as the spearhead of the 4th Armored Division and the US Third Army. He constantly managed to defeat German forces that had the advantage of superior armor and guns by consistently exploiting the relatively small advantages of speed and reliability of his vehicles. During the drive across France in 1944, Abrams forged a reputation as an aggressive tank commander noted for his concern for soldiers, his emphasis on combat readiness, and his insistence on personal integrity. LTG George Patton Jr. is often quoted saying of him, "I'm supposed to be the best tank commander in the Army but I have one peer, Abe Abrams. He's the world champion." Creighton Abrams was to play a decisive role in the drive to Bastogne.

COL Blanchard had been allowed to select his own route to the assembly area, which was located southwest of Bercheux. CCR column closed on this site alongside the Neufchateau-Bastogne road shortly before dawn on Christmas Day. Essentially attacking astride a road running generally to the northeast, the small villages along the way were used as objectives. Almost nothing was known of the German defenses along the 12-mile stretch of road to the American perimeter around Bastogne but tough resistance was expected. It was known though that capture of Bastogne was a top priority for the Germans and they were massing all available forces into this sector. Shortly after Christmas dawn, the battalions of CCR began their drive northward expecting strong German resistance to the north and a very real threat of a counterattack from the west.

The Americans were surprised to find light resistance in Vaux-les-Rosières and pushed on quickly to Remoiville. Here the Germans had placed the veteran 3d Battalion of the 14th *Fallschirmjäger* Regiment, which was taking advantage of the thick walls and cellars of the old town. LTC Abrams' 37th Tank Battalion was in the lead of CCR and paused on the hills overlooking Remoiville. Four battalions of artillery and a company of Sherman tanks blasted the town with rapid fire for ten minutes while A Team formed for an assault (Abrams had further task organized his companies, providing each with infantry support. These he called A Team, B Team, and C Team.) As soon as the fires lifted, A Team raced into the village with machine guns blazing, and once inside, the infantrymen dismounted from their halftracks and began clearing each building. The suppressing fire worked, keeping the Germans pinned down. When they tried to emerge to fight, they were cut down by heavy fire and hand grenades tossed into the cellars which quickly brought the survivors to the surface. The fight took most of the afternoon but by dusk, the CCR had taken 327 prisoners. Elements were pushed about 100 meters past the town but there was a large road crater where any bypass around it was impossible. It needed filled before the advance could continue.

That night, while the hole was filled, COL Blanchard assembled his commanders for a meeting. The plan he laid out was for an advance through Remi Champagne and Clochimont and turn to Sibret and the main Neufchâteau-Bastogne road. Heavy German defenses were reported at Sibret but there was promised air support to help deal with those. As usual, COL Blanchard mimicked the division commander in assigning an axis of advance to each of his maneuver commanders, allowing each to develop his plan to execute the mission. COL Blanchard had faith in his veteran commanders but still retained some oversight, particularly with the coordination of artillery fires and supporting units.

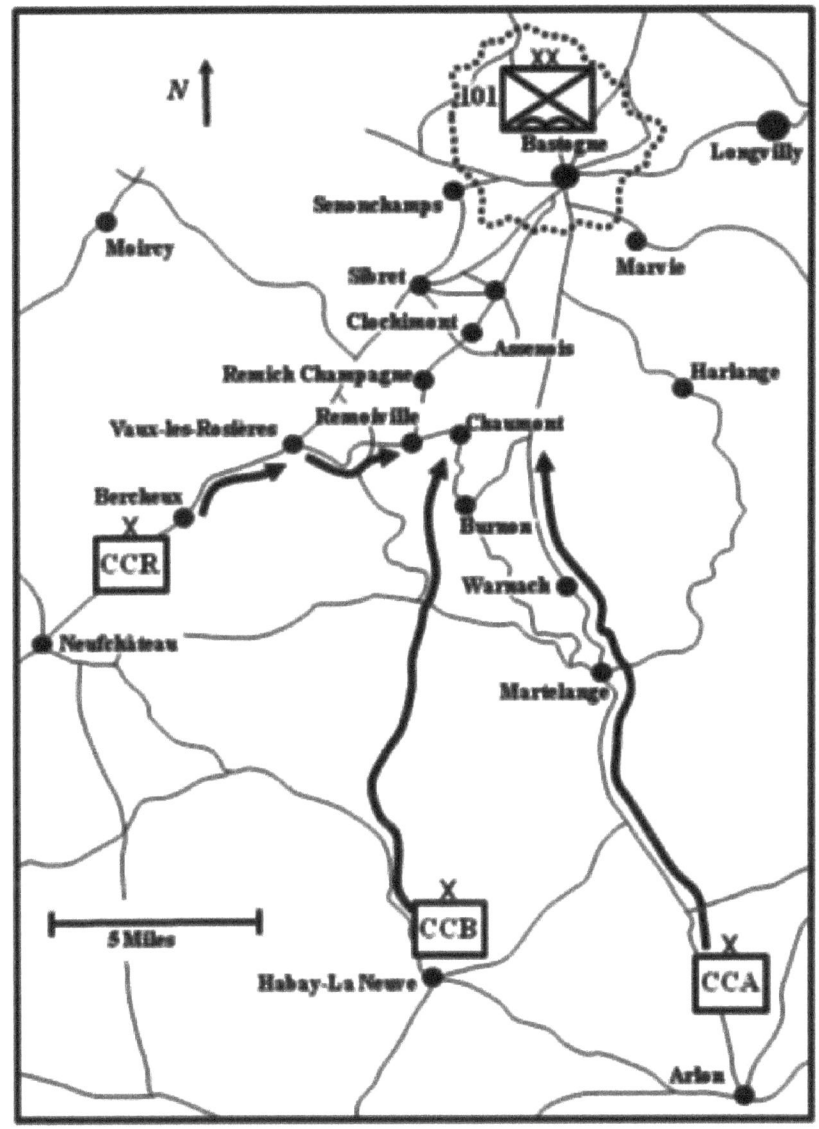

Map 2. The drive to Bastogne, 24-25 December, 1944.

This plan was put into operation on the morning of 26 December as the lead tanks started for Remi Champagne. The ground was frozen and the armored vehicles found it easy going even with the blanket of snow.

An unexpected but timely arrival of P-47 fighter bombers brought bombing and strafing runs into the town and surrounding woods, which dissipated enemy resistance quickly but when in closing in on Clochimont, the CCR carefully deployed in expectation of contact with the main line of enemy defense and a probable counterattack. These deployments were complete by 1500. By this time the 3d Tank Battalion was down to 20 Sherman tanks. Meanwhile the 53d Armored Infantry Battalion, weak to begin with, was short 230 men. Abrams and Jaques met on the road to discuss the next move. As they talked, dozens of C-47 transports streamed overhead toward Bastogne to deliver their cargoes. The need to reach the paratroopers of the 101st Airborne quickly was reinforced.

Abrams suggested they dash through Assenois and straight into Bastogne. LTC Jaques agreed. The town of Sibret was next on COL Blanchard's itinerary but it was known to be strongly held and would take time and effort to root the Germans out. At 1520, Abrams made a tough decision. While the bulk of CCR continued with the mission to reduce Sibret and other companies positioned to guard the division flank, two companies were detailed for a push straight to Bastogne. It was a risky move as there may not be enough troops to take Sibret and the two companies would be vulnerable to a German attack or they could get badly cut up by strong resistance further up the road.

LTC Abrams radioed his S-3, Captain William Dwight, and directed him to bring C Team forward. He also contacted the 94th Field Artillery Battalion through the liaison officer and updated them on his new plan. Abrams then asked that the 101st Airborne be alerted that American forces were approaching. CCR was alone in not having telephone wiring laid, so communications relied on the fickle radios. One result of this was that COL Blanchard was not yet told of Abrams' plan nor authorized such a move. Racing ahead of the general advance could invite confusion and his forces could get cut up by friendly air and artillery fire, not to mention fire from any Germans lying in wait, but Creighton Abrams was making the decision with the best information available at hand. With this move, he was still able to achieve the common purpose of defending the flank as well as achieve contact with the 101st Airborne Division. By 1620, the fire support plan was in place and C Team was in position.

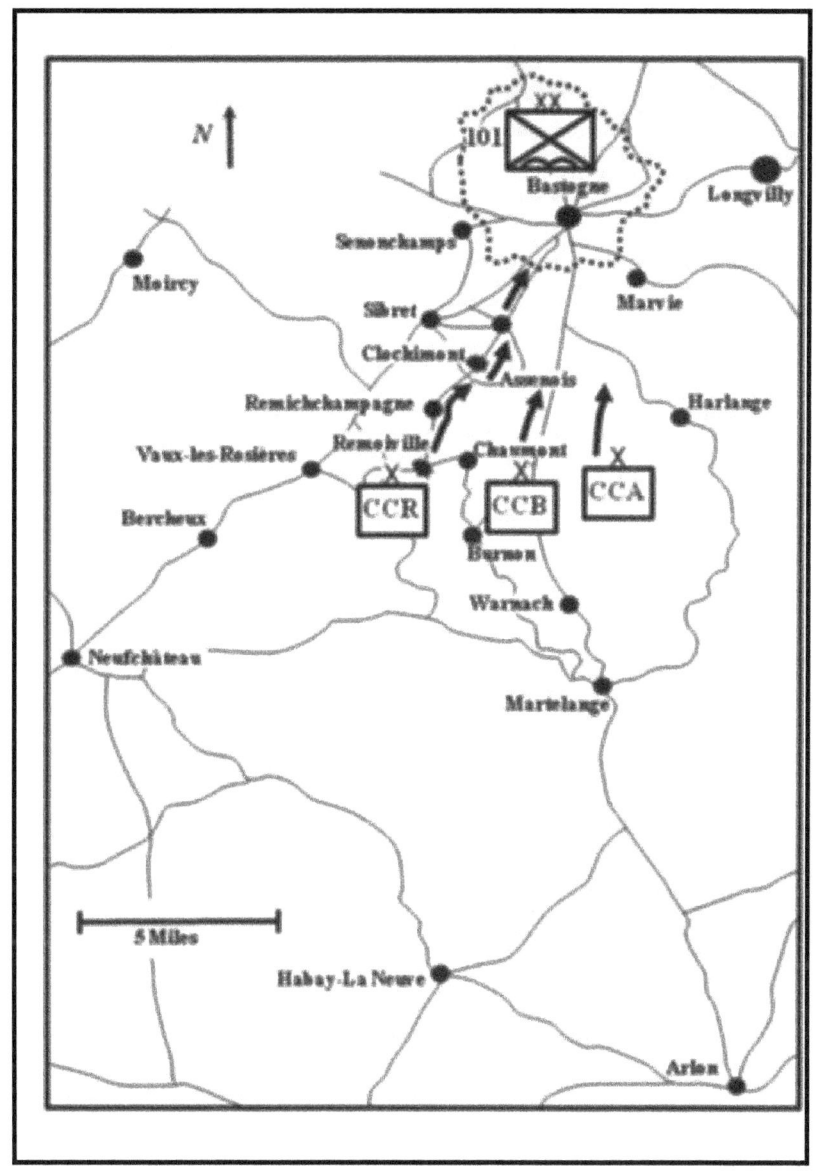

Map 3. The drive to Bastogne, 26 December 1944.

Abrams' C Team at this time consisted of the C companies of both the 37th and 53d battalions and Captain Dwight was now elevated to overall command. Dwight arranged the column with the Sherman tanks up front and with the halftracks behind and pressed up the road. Just when Assenois came into view of the lead tank, 10 volley-fires from 13 artillery batteries crashed into the center of town. Not waiting for the fires to lift, the tanks gunned their engines and they were soon beside the first buildings of the town with the mounted infantry close behind. The smoke and dust created by the shelling made the center of Assenois almost as dark as night. So close did the ground attack follow the artillery that not a hostile shot was fired at the tanks as they raced through the town's streets. This was quite fortunate as the Germans had posted eight deadly antitank guns to cover the road.

The American column pushed on into Assenois but at this point the attack faltered somewhat. Two tanks had made a wrong turn and a halftrack had fallen in with the tank column. Meanwhile the smoke and falling light of day reduced visibility to a few yards and added to the confusion. Additionally, the incoming supporting artillery fire rained shell fragments on the infantry riding in the open top halftracks. These men quickly dismounted to find the nearest shelter. Over 100 Germans from the 5th *Fallschirmjäger* and 26th Volks Grenadier divisions poured out of the cellars when the supporting fires finally lifted. The fight became that of savage close combat, much of it hand to hand.

Only five tanks and one halftrack emerged from the ongoing melee in Assenois to continue towards Bastogne. The rest of C Team was left battling for their lives in the streets. These tanks and the halftrack pushed on, spraying likely defensive positions along the road liberally with machine gun fire as they drove at top speed. The Germans though were able to quickly place mines in the road which disabled the halftrack. After clearing the mines and collecting the wounded the Sherman tanks continued on. At 1650 the lead tank spotted some American engineers up ahead preparing to assault a pillbox near the highway. These men turned out to be from the 326th Engineer Battalion of the 101st Airborne Division. To the surprise of the Americans, and undoubtedly the Germans too, the pillbox disintegrated as a tank round struck home. Contact was thus made with the beleaguered paratroopers. Twenty minutes later, LTC Abrams made his way to that point and shook hands with Brigadier General Anthony McAuliffe, the acting commander of the 101st Airborne Division, who had come to the outpost line to welcome the relieving force.

Fighting continued along the road to Bastogne for several more hours as the 53d Armored Infantry fought for possession of Assenois. By midnight of 26 December, the road was deemed safe enough to send 200 vehicles with badly needed supplies to Bastogne, as well as 22 ambulances to evacuate 652 seriously wounded soldiers there. With dash and élan the men of the CCR of the 4th Armored Division regained contact with the 101st Airborne Division. The threat of the loss of Bastogne was eliminated and the fate of the German offensive was sealed. For his actions, Creighton Abrams received his second Distinguished Service Cross.

For Further Reading

George Forty. *Fourth Armored Division in World War II*, Minneapolis, MN: Zenith Press, 2008.

S.L.A. Marshall. *Bastogne: The First Eight Days.* US Army in Action Series, Washington: Center of Military History, 1946.

Danny S. Parker. *Battle of the Bulge: Hitler's Ardennes Offensive, 1944–1945*, New York: Da Capo Press, 2004.

Lewis Sorley. *Thunderbolt: General Creighton Abrams and the Army of His Times*, New York: Simon and Schuster, 1992.

John Toland. *Battle: The Story of the Bulge*, Lincoln: University of Nebraska Press, 1999.

The Six Principles of Mission Command

1. Build Cohesive Teams through Mutual Trust
2. Create Shared Understanding
3. Provide a Clear Commander's Intent
4. Exercise Disciplined Initiative
5. Use Mission Orders
6. Accept Prudent Risk

Mission Command in the Bastogne case

1. Build Cohesive Teams through Mutual Trust. The division and combat command commanders had built cohesive teams through rigorous training prior to deployment into combat and reinforced these teams during the advance across France through the summer and autumn of 1944. At the lower echelons, the companies and battalions of the CCR had been task-organized often and had fought together many times. The mutual trust among the commanders is illustrated during operations as generally units were assigned axis of advance, giving the subordinate commanders the flexibility to formulate a plan of execution.

2. Create Shared Understanding. The men of the CCR, 4th Armored Division, had a shared understanding of the objective of the operation of reaching Bastogne and the embattled 101st Airborne Division. Each man knew the fate of the paratroopers depended on getting their quickly no matter what the Germans had planned. This in no small way motivated the commanders and the men of all arms to give their upmost effort. It was no secret that the road junction at Bastogne was key to the German plans and holding it would seriously hamper their offensive.

3. Provide Clear Commander's Intent. There is no doubt Creighton Abrams instilled in his men the need to reach Bastogne as quickly as possible and counted on everyone to do their part. Mission and fragmentary orders were short and concise but everyone knew his intent, what he wanted, and the speed in which he wanted it to happen. This command style was warranted due to the speed and distances experienced in mechanized warfare, especially when communications relied on relatively short range radio sets.

4. Exercise Disciplined Initiative. As the length of the flank increased so did the need for small unit leadership and initiative. At all levels men and units exercised discipline initiative in engaging German forces, by-passing when needed, clearing unexpected mines and obstacles, and main-

taining the momentum of the attack. Abrams certainly exercised initiative when he could not reach COL Blanchard by radio and decided to push on to Bastogne even though most of his unit was locked in bitter close urban combat. Abrams had served with COL Blanchard in combat for several months and knew he had his full support.

5. Use Mission Orders. COL Blanchard and LTC Abrams used mission orders throughout the operation. Orders were issued in conferences and unit visits when possible and via radio while on the move. The latter was necessary as the CCR did not have telephone capability. When communication was not possible the subordinate leaders made the best decisions they could, basing them on the situation at hand and remembering the commander's intent. The mission orders were broad in scope and encouraged the subordinate units to develop the situation in their assigned sectors and act accordingly in relation to the overall commander's intent. For example, LTC Abrams was told simply to attack along an axis of advance, clearing all German forces in order to clear a route to Bastogne. The tactics and methods to do that were not dictated to Abrams but left up to him.

6. Accept Prudent Risk. Creighton Abrams accepted prudent risk, weighing the merits of a quick drive to Bastogne against the lethality of the German defenses in his way. After consideration of the mission, the objective, and the enemy's likely situation and intent, he decided to push forward using speed to his advantage. It could well have ended in disaster with his unit chopped to pieces. German force dispositions were generally unknown along the axis of advance, and there was a possibility of deadly anti-tank fire at almost any point. Detaching a large part of his command while in contact was risky indeed, as Abrams may not have had enough forces to complete either mission. The need to relieve the American paratroopers from the German encirclement made this necessary. It proved to be the right decision.

Section 3: Cases at Company/Platoon/Squad Level

An Engineer Assault Team Crosses the Meuse, May 1940

Mark T. Gerges, Ph.D.

On 10 May 1940, the so called Phony War finally came to an end. The Germans invaded the Netherlands, Belgium, and France, seven months after the fall of Poland. French and British divisions rushed forward into previously neutral Belgium and occupied defensive positions along the Dyle River line. While large battles occurred along the expected invasion route in northern Belgium, creeping through the Ardennes forest was the true main effort of the German invasion. Led by Lieutenant General Heinz Guderian's XIX Panzer Corps (activated as the XIX Armee Korps and sometimes called the XIX Motorized Corps), the decisive operation of the German forces moved on five routes through the narrow forest roads unobserved and undisturbed by the allied air forces that focused on the anticipated main effort to the north. By midnight on 12 May, the lead elements of Guderian's forces arrived along the eastern bank of the Meuse River at the French city of Sedan, and throughout the night and early morning hours of 13 May, the main elements of the XIX Panzer Corps occupied their assembly areas.

The Meuse River was the main defensive obstacle anchoring the French defense along this sector. Sedan was the intersection between the Maginot Line defenses that ended just 20 miles to the south and the flank of the main Allied efforts in the Low Countries. Because it was the gap and not a critical threat itself, the Sedan sector was defended by second tier French reserve divisions. The French defenses were a patchwork of units under the command of the 55th Infantry Division, a "B" tier division consisting of reservists who served their active duty commitment in the 1920s, and for the last 15 years had done only the minimum of active duty training. Since mobilization the previous year, the division spent the winter and spring building bunkers to defend their sector along the Meuse River. Well sited, often constructed of reinforced concrete with machine guns or light cannons, the positions covered the most likely crossing points. Earth and wood reinforced bunkers covered the dead space between the more solidly constructed bunkers, with troops placed in intervals between the bunkers to protect the flanks. If anything, there were too many bunkers rather than too few. French soldiers spent more time as construction troops rather than training on their individual soldier tasks. Complicating the French defenses, the high command of the French decided after the opening of the campaign to reinforce the 55th Infantry Division with the 71st Infantry Division. On paper this looked simple and made sense with the division front of the 55th division reduced from 22 to 14 kilometers. However, the timing could not have been worse. Ordered to occur on the night of 13-14 May, much of the relief did not occur due to the German attack but still resulted in additional confusion as battalions prepared to displace just as the German attack commenced.

The German XIX Panzer Corps consisted of three Panzer divisions and a separate infantry regiment. The corps was to attack with three divisions abreast. The main effort was in the center of the corps sector with the 1st Panzer Division at the Gaulier factory, just west of Sedan. Reinforcing the division on their left flank was the *Grossdeutschland* Infantry Regiment, an elite unit crossing opposite the suburb of Torcey on the western bank. On the corps' right flank, the 2d Panzer Division prepared to cross near the village of Donchery and on the left most flank of the corps, opposite the village of Wadelincourt, was the 10th Panzer Division. The attack was to commence at 1500 on 13 May after five hours of bombardment of the French positions by German *Stuka* dive bombers.

The 10th Panzer Division planned to cross with two infantry regiments abreast, supported by the divisional engineer battalion. At this point the Meuse River is nearly 60 meters wide and too deep to ford. Other than near the city of Sedan itself, the 10th Panzer's approach to the river would be over 600 to 800 meters of flood plain that provided no cover or concealment. Fire support for the division was limited. In order to move rapidly though the Ardennes, the logistics train had been kept to a minimum, so artillery ammunition was extremely restricted. The 10th Panzer Division would only have their light 105mm howitzers available to support the crossing. Their heavy guns supported the XIX Corps artillery which, along with the *Luftwaffe's* JU-87 dive bombers, focused on the corps' decisive operation, the 1st Panzer Division crossing in the center. Compounding the problems facing the division was the narrow roads in the Ardennes which held up the engineer units with the inflatable rubber boats needed for the assault. One hour prior to the start time, no boats had arrived, and the engineer battalion commander rushed to the 10th Panzer Division's headquarters promising that the boats would arrive in time.

The division's engineer battalion, *Panzerpionier-Bataillon* 49 (49th Panzer Engineer Battalion) formed assault groups with teams of engineers and infantry squads that would cross in the initial assault to silence the French bunkers on the opposite bank. Twenty seven year old *Feldwebel* (Staff Sergeant) Walter Rubarth of the Second Company of the 49th Panzer

Engineer Battalion was assigned to lead one of these assault groups, consisting of five engineers supported by a squad of six infantrymen from the 1st Battalion, 86th Rifle Regiment. His orders were to cross the Meuse River south of a destroyed bridge and seize the bunkers on the opposite bank to support follow-on crossings by the infantry.

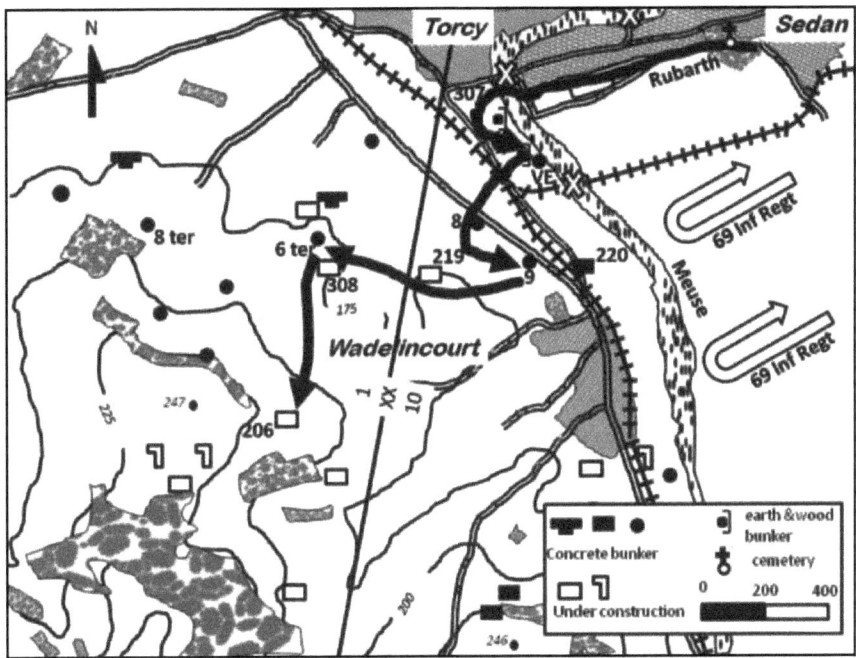

Maneuver of Rubarth's assault group.

At 1500, the last bombs from the JU-87 dive bombers hit the exposed French positions, German troops advanced from their assembly areas towards the river and the plan began to go wrong. It was immediately obvious that the Luftwaffe bombing had done little damage to the French artillery as accurate indirect fire fell on the now exposed German troops. In their haste to prepare their rubber boats for the crossing, the Germans were observed and heavy French fire landed among the boats and engineers. Of the 96 boats assigned to the crossing of the 86th Infantry Regiment, 81 were damaged and put out of action. The 86th Infantry Regiment was now unable to cross and only a single regiment, the 69th, had the equipment to serve as the division's decisive operation. Even that assault found itself stopped dead by heavy and accurate French fire before getting across the Meuse. All along the 10th Panzer Division's sector, the crossings were stopped at the near bank and the 10th Panzer Division was in danger of catastrophic failure.

The Meuse River at Sedan, taken in December 2002. This picture was taken standing at the crossing site of the 1st Panzer Division looking towards the 10th Panzer's crossing sites further south-east.

Author's collection.

Rubarth and his men had marched five kilometers that morning to reach their assault positions where Rubarth then went forward to observe the far bank. The French bunkers were easily seen and a German cannon was brought forward to provide direct fire support. As the last Stuka departed the area, *Feldwebel* Rubarth and his 10-man assault team moved from their assembly area at a cemetery in the eastern outskirts of Sedan, along the edge of town, through a sports field, and to the foundations of the blown bridge, concealed from observation by trees along the river. Rubarth's men had only two of the required three-man rubber boats. Rubarth ordered four men into the first boat and three in the other, leaving the rest of the squad behind to wait for more boats to become available but with all the Soldiers' equipment, the boats were dangerously overloaded and the water lapped over the gunwales. By mid-river the boats took fire and Rubarth ordered his machine gunner to fire at the nearest bunker's weapon slit. The machine gunner steadied his weapon on the shoulders of the man in front of him and returned fire while Rubarth's other men threw overboard their unnecessary equipment, including entrenching tools, to lighten the boats. Miraculously, Rubarth's group landed on the far bank with no casualties. The men landed near a strong earthen bunker, at which point French artillery began to fall on the crossing site. As far as Rubarth could see, he and his men were the only German forces to make it to the western bank and had no support from friendly units pinned-down on the opposite bank. Rubarth and Private First Class (*Gefreiter*) Podzus destroyed the nearest earthen bunker. Cutting through a wire obstacle, his men then attacked a second bunker, blowing a hole in the rear wall with a satchel charge and engaging the occupants with small arms and hand grenades. The French defenders surrendered and German troops on the far bank of the river began to cheer, encouraging Rubarth and his men. Continuing to move along a railroad embankment, then wading waist deep through a swamp, Rubarth's men destroyed two more earth and wood bunkers guarding the flank of a major concrete bunker, opening a hole of 300 meters in the French lines. Finally, coming to the road running behind the railroad embankment, Rubarth's squad received such heavy fire that they had to take cover. It was at this point that the *feldwebel* realized that he and his men were completely alone and unsupported on the enemy bank of the Meuse.

Bunker 220 along the railroad embankment at Sedan. *Feldwebel* Rubarth's actions just west of here destroyed the three earthen bunkers along the river protecting the left flank of this bunker, leaving it vulnerable when follow-on forces attacked it later in the day. Photo taken in December 2002.
Author's collection.

Out of ammunition, Rubarth returned to the crossing site. There he learned that a third boat with his remaining men had been hit and the four men killed. His platoon leader on the far bank organized men from other squads and rubber boats and four more engineers crossed the river to join Rubarth with additional satchel charges. While the four reinforcements brought him back to full strength, it was clear that the 10th Panzer's attack was stalled but Rubarth moved back to the railroad embankment and joined his men.

While machine gun fire pinned the Germans down at the railroad embankment, the French organized a local counterattack. The attack was beaten off but one non-commissioned officer was killed and two soldiers wounded. Reorganizing his remaining men, Rubarth crossed the tracks and attacked Bunker 8 and then Bunker 9 from the flank, increasing the breach in the French defenses. He then assaulted the strongest bunker in the area (Bunker 219), a partially completed concrete position with two 25mm guns located a little to the west. What Rubarth did not know was that his route, taken because of the French resistance, took him outside of the 10th Panzer Division's sector and into the 1st Panzer Division's. His attack on Bunkers 8 and 9, then Bunker 219 and its supporting positions, did nothing to help the 10th Panzer Division, his own unit, still stalled at the river. However Rubarth's actions were critical to the battle because they took his unit to the junction between the French first and second lines of defenses opposite Sedan. While he had destroyed a couple of bunkers in the first line opposite the 10th Panzer, his movement west cut perpendicularly across the French defense, opening a much larger hole and creating greater confusion among the French defenders.

Taking fire from their right flank, Rubarth and his men continued moving west where they took fire from bunkers 308 and 6B. Now inside the 1st Panzer Division's sector, Rubarth's actions were having unexpected results. The two battalions of the *Grossdeutschland* Regiment, delayed in house-to-house fighting in the village of Torcey, found the French fire on their positions lessening. Farther west, the 1st Rifle Regiment of the 1st Panzer Division penetrated the first line of French defenses and suddenly advanced into an area with reduced defensive fire due to Rubarth's actions, allowing them to consolidate on their objectives near dusk. Taking the three earthen bunkers along the river between 1500 and 1700 also had another unforeseen consequence. After 1700, another engineer team crossed the Meuse near the destroyed bridge, and attacked Bunker 220 from its now unprotected flank and moved south. This attack undermined the French defense against the 10th Panzer, allowing the division to cross later that evening.

By dark, Rubarth's 11-man team had suffered six dead and three wounded. They consolidated on the heights above the village of

Wadelincourt about a kilometer west of the river, linking up with another engineer platoon from their battalion after nightfall. *Feldwebel* Walter Rubarth received a battlefield promotion to lieutenant and was awarded the Knight's Cross for his actions on 13 May 1940. After the French campaign, he served in the 1941 invasion of Russia and was killed outside of Gzhatsk, Russia on 26 October 1941.

For Further Reading:

Robert A. Doughty. *The Breaking Point: Sedan and the Fall of France, 1940*. Hamden, CT: Archon Books, 1990.

Karl-Heinz Frieser. *The Blitzkrieg Legend: The 1940 Campaign in the West*. Annapolis, MD: The Naval Institute Press, 2005.

Alistair Horne. *To Lose a Battle: France 1940*. Reading, UK: Cox and Wymann, Ltd., 1969.

The Six Principles of Mission Command

1. Build Cohesive Teams through Mutual Trust

2. Create Shared Understanding

3. Provide a Clear Commander's Intent

4. Exercise Disciplined Initiative

5. Use Mission Orders

6. Accept Prudent Risk

Mission Command in the Engineer Assault Team Crosses the Meuse case

1. Build Cohesive Teams through Mutual Trust. Despite the fact the *Feldwebel* Rubarth's assault team was a temporary grouping of an engineer squad and infantry squad for this particular mission, the high level of training allowed them to operate effectively. All units in the XIX Panzer Corps could be considered elite due to their combat experience in Poland and extensive training since the Polish campaign. This experience led to a high degree of cohesion and trust among the men.

2. Create Shared Understanding. During preparation for the 1940 Campaign, the German Army planned carefully and extensively. The XIX Panzer Corps and its subordinate organizations rehearsed the mission to cross the Meuse numerous times in the months leading up to the operation. The command philosophy and practice at all levels in the German Army was similar to what the US Army now calls Mission Command with a major emphasis on giving small units objectives but little other specific guidance. Officers and NCOs like Rubarth were expected to solve tactical problems on their own using their resources and initiative. In the case of the river crossing, the task—an assault river crossing in the face of determined resistance – placed great importance on small unit leaders finding ways to reach the far bank of the river and assault the enemy positions so that a penetration was possible. Rubarth's squad clearly understood this.

3. Provide a Clear Commander's Intent. While the actual orders issued to Rubarth are lost, the intent and verbal instructions remain—cross the river and assault the bunkers, which provided clear intent to the soldiers under his leadership.

4. Exercise Disciplined Initiative. Rubarth understood his orders and the intent, and evaluated these orders on the basis of the situation which actually existed when he crossed the Meuse. Specific orders to attack particular bunkers within his division's sector may have led to disaster

because Rubarth moved along the route of French vulnerabilities which just happened to lead out of his division's sector into a neighboring division's. Lavish adherence to a plan written without the knowledge of the actual layout of the French defense would have led Rubarth's men into the beaten zone of the French bunkers. Instead, Rubarth's initiative combined with his understanding of commander's intent led to his squad playing a pivotal role in the successful German crossing of the Meuse River.

5. Use Mission Orders. As noted above, the XIX Panzer Corps and its subordinate organizations rehearsed the mission to cross the Meuse numerous times in the months leading up to the campaign. The orders for the crossing were broad and maximized freedom of action for subordinate units. Divisions were assigned sectors rather than specific objectives and timetables. Indeed, the entire corps operations order, including the fire support annex, was only nine pages long. The execution order by the corps commander upon arriving at the Meuse told his subordinate commanders to cross the river and capture their objectives according to the map exercises! Rubarth's orders allowed him to figure out how to accomplish his mission, including the route to take and bunkers to attack. He was to cross the river, assess the situation, and by understanding the division's broader mission (cross the Meuse), do what he deemed appropriate at his level to support his higher command's mission.

6. Accept Prudent Risk. *Feldwebel* Rubarth balanced the risk to his men with the risk to the division's decisive operation while understanding the importance of accomplishing his mission. Even when it appeared that he and his men were the only German forces across the river, he understood that he must accomplish his mission with the forces available to him. Still, in choosing his route of advance, he moved his squad along the path of least French resistance, rather than take on the French bunkers in a frontal assault. By mitigating risk in this way, he actually was able to create opportunities for his own soldiers as well as the units on the far side of the river waiting to cross.

Capturing Eben-Emael

The Key to the Low Countries

Nicholas A. Murray, Ph.D.

Fall Gelb, the German plan for the invasion of France and the Low Countries called for Army Group B, Sixth Army in particular, to quickly drive through Holland and Belgium in order to help fix the Allied forces in place. This was in order to facilitate the German main effort to the south, the *Sichelschnitt*, which had as its aim the cutting off of the core of Allied armies in order to destroy them in a large pocket. The Belgian fortress of Eben-Emael lay on the axis of attack of Sixth Army (Map 1). The fortress covered several key bridges across the Meuse-Albert Canal just to the west of Maastricht. The capture of the bridges was crucial to the success of the German invasion. Eben-Emael was considered by many to be the most powerful fortress in the world and it needed to be taken quickly if the German plan was to work. In the first ever glider assault, elements of the 7th *Flieger* Division rapidly captured the fort, opening the way to the west for Sixth Army. The success of the mission came about as a direct result of the flexibility, personal initiative, cohesion, and innovation of the Soldiers and their commanders.

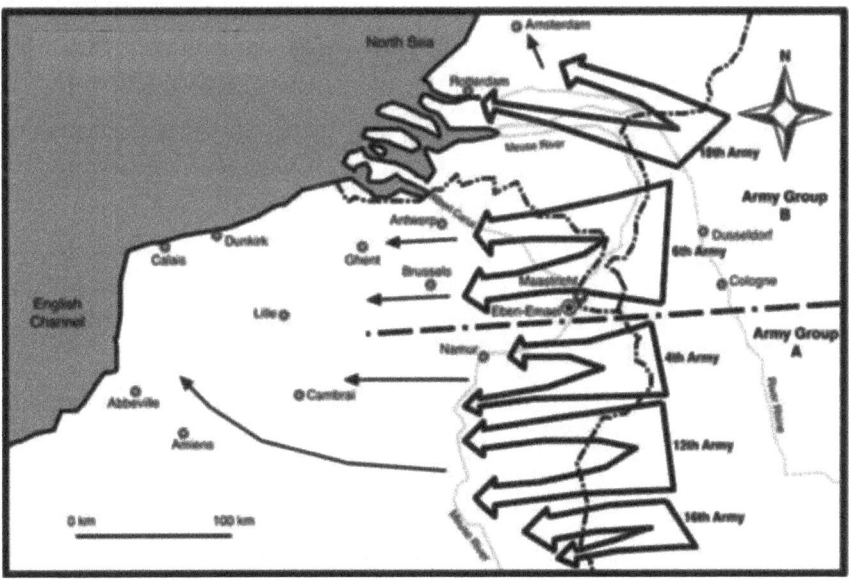

Map 1. The German plan to invade France and the Low Countries.

In October 1939 Hitler added an order to *Fall Gelb* for the capture, by paratroopers, of Eben Emael to assist in maintaining the high operational tempo demanded by the plan. GEN Karl Student, commander of 7th *Flieger* Division, was tasked with the mission. He allocated CPT Walter Koch as overall commander for the capture of the bridges and LT Rudolf Witzig with the specific task of capturing Eben-Emael itself. Witzig was chosen because he was an excellent officer as well as the fact that he was the commander of the only paratrooper assault engineer unit, known as *Sturmgruppe Granit*, a unit whose skills were essential for an attack upon a fortress. Tasked with his mission, Witzig began to train his men.

Initially, it was thought that there would only be a week or so to train the soldiers chosen for the mission. Despite this, Witzig was confident in the skills of his men and trusted them to do well. As *Fall Gelb* was delayed until the spring of 1940, Witzig was able to more thoroughly drill his men over the six months gap. He oversaw all the aspects of training with CPT Koch occasionally showing up to check that all was well. Koch left Witzig to train the men as he saw fit. Koch trusted Witzig and knew that he was the expert in this type of assault. Thus Koch largely limited his role to support, providing additional troops for the mission when it became clear that Witzig's platoon was not large enough, and facilitating the platoon's training without overly interfering. The long period of training reinforced the already high cohesion of the unit and it allowed for the soldiers to practice for a variety of scenarios. This provided them with great flexibility for their mission.

Witzig was also largely responsible for design of the tactical plan of attack and he worked closely with his senior NCOs to accomplish this. He was helped by clear mission orders:

> Capture by surprise the surface of Eben-Emael. To guarantee the transit of the Army over the Meuse-Albert Canal, neutralize the artillery and anti-aircraft casemates and turrets. Break any enemy resistance and hold until relieved.

Additionally, Witzig pointed out that the process of working with his subordinates was reinforced by the unit's "trust and loyalty from bottom to top and from top to bottom." This cohesion was to pay great dividends during the attack itself.

The attack force was arranged in 11 squads of seven or eight men each for a total of 83 men and two officers. The small size of the squads was largely a product of the technical specifications of the *DFS* 230 assault gliders in which the unit was going to land on the roof of the fort. Gliders were chosen because their lack of engines (less noise) meant that surprise was more likely. They had the added advantage of being able to land on a precision target, something that parachuting onto the target could not guarantee. There was thought to be too great a risk of a loss of surprise if parachutists took time to concentrate before assaulting their targets. To try and guarantee that the assault force had at least one officer available, LT Egon Delica was to land with the First Glider and Witzig with the Eleventh (Reserve) Glider. The 83 other ranks contained 28 NCOs, who were to prove crucial to the success of the mission.

Each squad was allocated one main target with the idea that once it was neutralized they should then start engaging subsidiary positions. As such, this meant that each glider would land separately from the others and the men would thus have to act largely on their own initiative, at least at first. Once the main positions had been dealt with, the Germans were to establish contact with the other elements of Koch's force (attacking the bridges themselves), and coordinate with the lead elements of Sixth Army to facilitate the river crossing.

The attack was scheduled to be one of the first operations for the German invasion of the west on 10 May 1940. Despite six months of training and practice, however, things went wrong shortly after the troops were airborne. Two of the gliders, which were being towed by transport aircraft, lost their tows and had to make emergency landings short of the target: second squad under the command of SGT Max Maier and eleventh squad (Witzig's). Thus, the operation lost its commanding officer and a senior NCO before it arrived.

It is worth looking at both Maier's and Witzig's reaction to this setback. Once the tow rope was lost, thinking quickly, Witzig ordered the glider pilot to look for a field where they could not only land but from where they could be re-lifted. This done, as soon as they were on the ground, Witzig ordered his men to clear a temporary runway and himself set off to find the nearest German transport airbase. There, he commandeered a transport plane and flew to collect his glider before heading off to land several hours late at Eben-Emael. SGT Maier was equally intrepid. He hitched a ride on a motorbike, with his squad's CPL P. Meier, to the nearest town. There he commandeered two small cars to transport his squad to Eben-Emael. Unable to take the cars further than the Meuse River, he elected to continue on foot. Maier persuaded some German engineers to put his men in one of the first boats to cross the river and from there they made their way (via another commandeered vehicle) to the bridge nearest to Eben-Emael. This had been demolished by the Belgians. Undaunted

Maier attempted to cross the Albert Canal alone to reconnoiter a way forward but was killed in the process. CPL Meier now took charge and after waiting a few minutes, he crossed the canal. From there he stole a bicycle and rode to the fortress. However, he was unable physically to get to the rest of *Sturmgruppe Granit* (by this time on top of Eben-Emael itself) and instead scouted along the moat on the northwestern side of the fortress before making contact with SGT Haug of fifth squad. Imparting his story by shouting, he let Haug know that he would attempt to get his men to the fort as quickly as he could. Heading back to rejoin his men, and still unable to reach *Sturmgruppe Granit*, Meier instead linked up with relieving German troops from Sixth Army and aided them in crossing the Albert canal. Whilst that was going on, each of the nine other gliders had landed on the fort and the Soldiers had set about their mission.

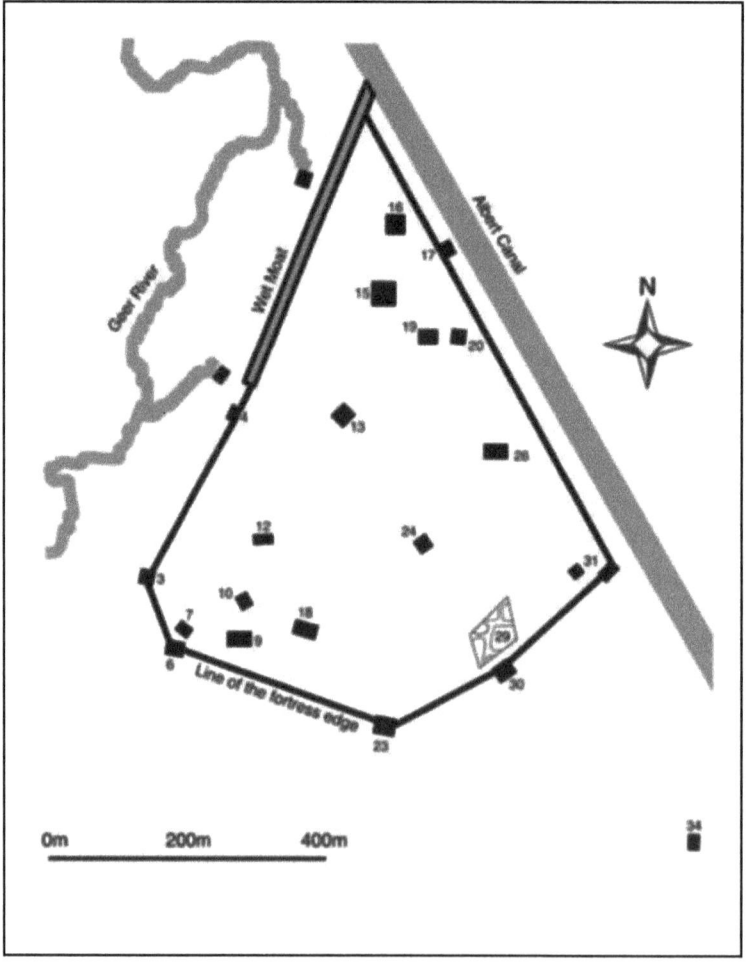

Map 2. Fortress Eben-Emael casemates with armored cupolas indicated by number designation.

The first nine squads each had targets designated for when they landed, with the tenth and eleventh squads acting as the reserve. Given the failure of second and eleventh squads to show up, the situation was a little more complicated. Ideally LT Delica, who was senior on the ground, should have taken charge. His glider, however, had landed a fair way to the south and his squad was busy dealing with its target position, a 75mm-gun casemate (18, Map 2). Unable to contact Witzig or Delica, SGT Helmut Wenzel of fourth squad took command and established headquarters for *Sturmgruppe Granit* inside the machine-gun casemate (19, Map 2), which his men had captured minutes earlier. SGT Wenzel (described by LT Witzig as "a first rate man, an old engineer with vast experience, a vigorous troop leader") was fully familiar with the mission and continued with the plan. He had his radioman establish contact with Koch in order to inform the overall commander when his men had taken their main objectives, as well as to gain situational awareness as to the whereabouts of the relieving troops. Meanwhile the other squads had landed in proximity to their targets and set about dealing with them. As Witzig described it, "they didn't need to ask questions. They had their orders, and they did them."

Squad eight, under SGT Unger, had been the first to land and they immediately set about dealing with their target, a 75mm-gun position in an armored cupola (31, Map 2). Being the first to land, they came under a hail of small-arms fire and they took their first casualty almost immediately. They close-assaulted a Belgian position which threatened their advance of the main objective. They did so with the help of fifth squad which had already successfully disposed of their own targets: the anti-aircraft guns (29, Map 2) and the armored observation cupola (30, Map 2). SGT Haug, of fifth squad, had used his initiative to support eighth squad's attack. This was a crucial and timely intervention as the Belgian Soldiers inside the cupola had just loaded two 75mm rounds when the German engineers detonated their shaped charge knocking out the position. The Germans then blew in the armored access doors of the gun position effectively forcing the Belgians to abandon it.

Separately, squad three, under SGT Arendt, had an unanticipated difficulty in their attack on another of the 75mm-gun casemates (12, Map 2). Their plan anticipated an access door to the casemate, or possibly an armored observation cupola, through which they might neutralize the casemate's garrison. However, there was no obvious door and no observation cupola. SGT Arendt decided to improvise. He detailed his men to blow in one of the gun mounts. This was a difficult task as the charges were not designed for this and neither had his men trained to do it. They managed, however, to place a charge in one of the gun embrasures before detonating it, destroying one of the 75mm guns. This also put out of action the casemate, into which they could now enter. Rather than stop with the neutralization of the casemate, SGT Arendt entered and dropped explosives down a connecting stairwell. This had the effect of deterring any Belgian defenders from seeking to re-occupy this casemate.

The targets on the fortress were dealt with in a fashion similar to the above, with the exception of the 120mm-gun cupola (24, Map 2), and a 75mm-gun casemate (26, Map 2). These had been the targets of squad two, which had not arrived. This presented a problem. The 120mm-gun cupola contained the most dangerous weapons on the fortress and it was essential that they were dealt with. Glider pilot Heiner Lange of fifth squad realized the danger and took it upon himself to knock out the cupola. Although he was not an actual engineer, he had trained with his compatriots and was familiar with how the explosives worked. Despite being wounded, he succeeded in detonating his charge. This, however, did not completely succeed in knocking out the position. It only partially damaged the guns and the Belgian crew re-occupied the position until fifth squad again attacked it and eventually put it permanently out of action. These actions show great initiative, as this was not a part of the plan for fifth squad. This left one of second squad's positions to deal with. SGT Wenzel, in command in the absence of Witzig, ordered tenth squad, under SGT Hübel, to attack the position and it was put out of action relatively promptly.

Although not all of the Belgian positions had been completely destroyed, the fort had been sufficiently neutralized for Wenzel to signal Koch that the mission was accomplished. All that remained was to make sure the fortress stayed relatively quiet. No small undertaking with such a small force. Witzig's glider landed on the fort at around 0800, about three to four hours after the start of the action. He liaised with Wenzel and discovered that their relief by 4th Panzer Division, originally expected at 1000 hours, was hours behind schedule. Witzig now had to deal with Belgian counter-attacks without the expected support. Witzig organized his men to set up a defense of the surface of the fort and called in the help of the Luftwaffe. This is where the German's superior combined arms really counted. Belgian troops moving towards German positions on Eben-Emael were subjected to numerous air raids which both slowed them down and reduced their combat power. Despite this, Belgian troops made several attempts to clear the roof of the fort and to re-occupy their gun positions. Ultimately these attempts failed to alter the outcome. German initiative and combined arms, along with the dogged tenacity of the German engineers, would allow Witzig's men to hold on overnight. That being said, his position was serious. His men were short of ammunition, explosives, and most importantly drinking water. Witzig was not sure how much longer his men could hold.

The Albert Canal as seen from Fort Eben Emael.

US Army photo.

In order to facilitate the relieving troops crossing the Albert Canal, several troops from eighth squad swam across the canal to act as guides for the reinforcements. The subsequent attempts to cross the canal by relieving troops from the 51st Pioneer Battalion came to naught as a machine-gun and anti-tank gun casemate (17, Map 2) covered the canal and could not easily be attacked (it was tucked down by the edge of the canal, a fair way below the level of the fort). The steep walls of the cut, through which the canal went, prohibited easy access to it. Witzig's men improvised by lowering explosive charges down and shoveling dirt to block the casemate's observation slits. Although this did not stop the Belgian machine-gun fire, it did reduce its effect. However, the relieving pioneers in rubber boats were still unable to get across the canal in daylight.

SGT Portsteffen, of the 51st Pioneer Battalion, had made several attempts to cross the canal with his men during the late afternoon and evening. All had failed. He decided to wait until dark before again attempting a sortie. This time he and his men were successful. Regrouping on the western side of the canal, they worked their way up the dry moat on the northeast side of Eben-Emael to casemate 4 (Map 2). This casemate had unsuccessfully been attacked several times by Witzig's men, who needed to neutralize it to open a route for the relieving forces. At casemate 4, Portsteffen personally used a flamethrower to suppress the bunker before his men neutralized it with explosives. This opened a route over the casemate and up the steep side of the fortress to the surface. Portsteffen, alone, ascended the side of the fortress eventually meeting with men from ninth squad. Witzig joined them at around 0700 on 11 May. *Sturmgruppe Granit* had held its position for almost 24 hours longer than anticipated. Now out of water and short of ammunition, the relief was just in time.

Portion of Fort Eben Emael.

US Army photo.

Relieving *Sturmgruppe Granit*, the men of 51st Pioneer Battalion, guided by Portsteffen, proceeded to reduce the remaining Belgian positions. By mid-morning they had forced most of the defenders into the interior of the fort. With more forces and combat power arriving, the Germans were now in almost complete control of the surface of the fort. Still, however, there was desultory resistance. This ended at 1215 when the Belgian garrison surrendered. They had suffered 88 casualties inside the fort and around 1,000 prisoners fell into German hands. Of the 85 men of *Sturmgruppe Granit*, 24 were killed or wounded (28% of the force). Despite this high casualty rate, the fact that the glider troops accomplished their mission demonstrated the high level of trust and cohesion among them. This operation was also an enormous propaganda coup for the Germans. They had tried something never before attempted in war and they had pulled it off. After the operation, junior officers and NCOs were prominent among those receiving awards for outstanding leadership and initiative at Eben-Emael.

The capture of Eben-Emael opened the route across the Albert Canal and through central Belgium. The German Blitzkrieg invasion of France and the Low Countries would likely not have proceeded as well without the success of this daring mission. The exercise of disciplined initiative by the junior leaders was absolutely critical to the success of the mission.

For Further Reading

Simon Dunstan. *Fort Eben-Emael: the key to Hitler's victory in the West* (Oxford, UK: Osprey Publishing) 2005.

James E. Mrazek. *The Fall of Eben-Emael* (Novato, CA: Presidio Press) 1999.

Tim Saunders. *Fort Eben-Emael* (Barnsley, UK: Pen & Sword Books Ltd.) 2005.

Gilberto Villahermosa. *Hitler's Paratrooper: the life and battles of Rudolf Witzig* (London, UK: Frontline Press) 2010.

The Six Principles of Mission Command

1. Build Cohesive Teams through Mutual Trust

2. Create Shared Understanding

3. Provide a Clear Commander's Intent

4. Exercise Disciplined Initiative

5. Use Mission Orders

6. Accept Prudent Risk

Principles of Mission Command in the Eben-Emael case

1. Build Cohesive Teams through Mutual Trust. The officers and men were the only paratrooper engineers in the German forces, and they spent around six months together training specifically for the assault on Eben-Emael. After the operation, Witzig pointed out that the process of working with his subordinates was reinforced by the unit's "trust and loyalty from bottom to top and from top to bottom."

2. Create Shared Understanding. The long time the men spent together in training for the mission meant that they all knew what they were required to do, and could see that everyone was up to the task in hand. Furthermore, their task was clearly set out for them.

3. Provide a Clear Commander's Intent. In the operational order, the mission was very simply and clearly set out:

"Capture by surprise the surface of Eben-Emael. To guarantee the transit of the Army over the Meuse-Albert Canal, neutralize the artillery and anti-aircraft casemates and turrets. Break any enemy resistance and hold until relieved."

This easily translated into simple but clear instructions for each of the unit's teams on the mission itself. While this allowed the unit to create its own detailed plan to capture the fort, it also allowed the freedom of action if that plan did not go smoothly.

4. Exercise Disciplined Initiative. The paratrooper engineers who attacked the fort were highly adept at the squad and platoon level and possessed great confidence in their leaders. They had regularly practiced attacking fortifications and they were practiced with their equipment and training. Although the mission was given to the regiment itself, its commander passed on the task to CPT Koch. He periodically oversaw training for the attack, and in turn largely deferred to LT Witzig who was charged with carrying out the mission. Thus, initiative was devolved to

the lowest appropriate level. Additionally, the Soldiers of 51st Pioneer Battalion that linked up with *Sturmgruppe Granit* also displayed exceptional initiative, particularly Sergeant Portsteffen whose men were the first pioneers to assault across the canal to relieve the airborne troops at Eben-Emael.

5. Use Mission Orders. Although formal orders were given prior to the assault on Eben-Emael, they were fairly broad and maximized freedom of action for subordinate units. Individual squads were assigned enemy positions to attack, with the understanding that the mission required flexibility. To that end, each glider contained a considerable amount of engineering equipment over that required for its specific role. It is worth noting, LT Witzig's personal glider as well as another, were delayed. However, despite lacking their assigned leader and effective communication, SGT Wenzel took charge and proceeded to adapt to the situation at hand. The setback of two gliders failing initially to show up did not prevent the troops successfully conducting the mission without their commander. To a large extent this success is explained by LT Witzig himself who described SGT Wenzel as "a first rate man, an old engineer with vast experience, a vigorous troop leader." LT Witzig went on to describe why the men were so successful despite the loss of their senior leaders: "they didn't need to ask questions. They had their orders, and they did them."

6. Accept Prudent Risk. The use of the first ever glider attack was a significant risk. However, the sheer surprise gained mitigated against a more effective enemy reaction. In addition, the immense importance of Eben-Emael meant that a high degree of risk was acceptable in relation to the potential gain from success. If the mission failed there was a significant risk of very high casualties as the main force attempted a forced river crossing under heavy defensive artillery fire.

The Bridge at Mayenne, France 1944

Kevin M. Hymel

LTG George S. Patton, Jr. needed to capture the French town of Mayenne if he was to destroy the German Seventh Army. In early August of 1944, Patton's newly activated Third Army was on the move, but with most of his armor heading west into Brittany, the capture of the town and its three bridges over the Mayenne River afforded him an excellent location to begin an eastern encirclement with the British assisting by pushing south from Caen. If the Germans managed to destroy the three bridges, however, they could stall the encirclement and safely escape. To capture these vital bridges intact, Patton chose BG Raymond S. McLain's 90th Infantry Division (ID).

McLain had commanded the unit for less than 10 days. The 90th ID had been considered a problem division since landing in Normandy on D-Day where it initially performed poorly against the Germans. The unit's crest, "T" and "O," stood for "Texas and Oklahoma," the home of the division's officers in World War I. The men however preferred the letters to stand for "Tough 'Ombres." McLain intended the unit to live up to the aggressive title. By the end of his second day in command, he had relieved 16 officers. BG William Weaver, a no-nonsense warrior who led by example, was assigned as McLain's assistant division commander.

With the Germans all over Patton's front reeling in confusion, McClain took immediate advantage of the situation. He organized Task Force (TF) Weaver, under the command of BG William Weaver, to serve as a fast hard-hitting mobile strike force that would exploit any penetration of enemy lines. The task force was composed around COL George B. Barth's 357th Infantry Regiment. McClain increased its strength by adding the 712th Tank Battalion, a company from the 607th Tank Destroyer Battalion, and the 345th Field Artillery Battalion to provide fire support. In addition, the 90th Reconnaissance Troop would serve as the TF's eyes and the 315th Engineer Battalion would clear mines and obstacles. A squadron of Republic P-47 Thunderbolt fighter bombers provided air support while a company from the 315th Medical Battalion cared for the wounded.

Late on the night of 4 August, COL Barth was called to division headquarters. After being informed that his infantry regiment was to be part of TF Weaver, Barth was given his mission. "The force was ordered to drive south and seize Mayenne, a big town 30 miles behind the enemy front line," Barth recalled. "It was thought that there was only light resistance at the point we were ordered to break through."

As the sun rose on 5 August, the task force headed out from St. Hilaire with the armor and reconnaissance vehicles screening up front while infantry-laden trucks followed close behind. Other elements followed, averaging 20 miles an hour. People lining the streets threw flowers at the advancing soldiers. When the columns slowed, the locals ran forward and offered bottles of wine. Members of the French Resistance also approached offering information on enemy positions.

"The men seemed to sense the fact that something big was in the wind," reported COL Barth. "An undercurrent of excitement seemed to go down the column and you could almost see the men's spirits rise. Morale was on the way up." The task force reached the outskirts of Mayenne around 1430 where two reconnaissance vehicles suddenly blew up at a mined roadblock. From the woods on both sides of the road, German infantry opened fire with machine guns and anti-tank weapons. Barth rushed forward through enemy fire to direct his mortar teams, and then directed a company-level assault on the German position. The men quickly took the roadblock. The battle to capture Mayenne had begun.

BG Weaver, scouting up front with a carbine in hand, led the task force into the city. "I remember General Weaver moving out in front of us," recalled 1LT Burrows Stevens, the commander of Company B. "I had to run to keep up with him." Weaver decided on a two-pronged attack. He ordered MAJ Edward Hamilton to attack the bridge in the center of the city with his 1st Battalion. Weaver then ordered Barth to dismount the rest of the regiment from the trucks, envelop the town from the south, and cut off roads leading out of Mayenne to the east. Hamilton pushed his men and tanks forward, encountering only slight resistance. To ensure no surprises, his artillery and tanks hit possible enemy positions on the high ground on the western edge of the city. Hamilton's men quickly captured the western section of the city only to discover that Germans had destroyed the two southern bridges spanning the river and had rigged the third for demolition.

1LT Richard Smith, a mortar section leader, shot the locks off a door to a house on the riverfront. After running to the attic he observed that the bridge was rigged for destruction with eight 500-pound aerial bombs. On its eastern edge, the Germans had posted two 88 mm guns, a 20 mm anti-aircraft gun, and several tanks. Using his new perch as an observation post, he called for and adjusted mortar fire to drive away one of the enemy gun crews.

As the Americans pondered the best way to capture the bridge, they came under fire from the Germans on the east bank. Unexpectedly, a

German tank on the eastern side of the river approached from the north, heading for the American positions. Hamilton called artillery on the tank and ordered his anti-tank platoon north to block any enemy attempts at flanking his left. As the one anti-tank team repositioned their gun to fire on the tank, German artillery opened up, killing one man and wounding two others but the team was still able to fire their anti-tank gun, dispatching the German tank.

Sherman tanks then took up positions along the river and began firing at the Germans on the other side. Behind US lines on the west bank, a lone German vehicle pulled up intending to cross the bridge. The occupants did not realize they were now in American territory. A single round from a tank destroyer at point-blank range ended the vehicle's journey. "The result," reported Hamilton, "was carnage." When exchange fire hit a French civilian, SSG Charlie Lancaster broke cover, raced over to the man and carried him to safety.

From a corner building near the bridge, MAJ Hamilton issued his plan of action. "It was simple and straight," recalled 1LT Stevens. Hamilton planned to hit the far bank with a 10-minute artillery barrage. As soon as it ended, Stevens' Company B would rush across the bridge followed by engineers who would disable the eight bombs. Finally, tanks would cross to support the infantry and engineers. Machine gunners would provide covering fire from buildings along the river and along a wall on the west bank.

As planned, the artillery began pounding away at the German positions at 1750. One round hit a stockpile of 88 mm ammunition sending a huge blast into the air and a pall of smoke over the attack route. Realizing the smoke would obscure the enemy's view of the bridge, Hamilton ordered a halt to the artillery and told 1LT Stevens to immediately charge the bridge. At that moment, the plan fell apart. The lead infantry squad froze. Only a lone tank, commanded by 1LT Charley Lombardi, clanked over the bridge with its cannon firing.

Seeing this disaster in the making, 1LT Stevens called, "Follow me!" and ran out behind Lombardi's tank. He followed it over the bridge emptying his only weapon, a German Walther P-38 pistol. Inspired by Stevens' courage, the platoon and two engineers followed. Along with a hail of machine gun and small arms fire, the enemy fired rounds from the 20 mm anti-aircraft gun down the length of the bridge. One shell tore into engineer James McCracken and, as Hamilton recalled, he "momentarily seemed to disappear." The round that killed McCracken also tore the leg off another engineer.

Without engineers, the infantrymen cut some of the wires leading to the bombs as they moved toward the opposite bank. SSG Raymon Lopez led the men to the row homes along the eastern bank where they tossed grenades into cellars and fired their rifles at every window. Lombardi's tank rolled off the bridge and onto a side street where it engaged a German gun and its crew and then moved up onto the high ground at the far eastern side of the town.

1LT Stevens' men found themselves alone on the eastern bank. The platoon directed to follow them had not crossed the bridge. Instead of supporting Stevens' unit, some of these men scrambled into doorways along the street for protection. As MAJ Hamilton charged down the street ordering the men to fire their weapons, 1LT Smith began pushing the men out of their hiding places. While this was going on, Stevens returned to the bridge, disabled the remaining German demolitions, and crossed to the west side to look for the missing platoon. Finding the lieutenant in charge paralyzed with fear, Stevens ordered the platoon sergeant to get his men moving. Stevens then assured the cowering soldiers that the first squad had crossed with no casualties (he had passed McCracken's lifeless body on his return trip and knew this was not the case) and that all the Germans next to the river had been cleared out.

Stevens' encouragement, along with cajoling from other officers and NCOs, finally got the platoon moving toward the bridge. A platoon of tanks followed. As the men began to run, a German vehicle appeared behind them, also moving in the direction of the bridge. One soldier halted it with a single round from a bazooka. It turned out to be an ambulance packed with 12 wounded men. Miraculously, none were hurt from the round.

The men then stormed the bridge under fire. An enemy bullet bounced off a sergeant's helmet and killed the soldier beside him. Stevens stopped along the way and picked up the M-1 rifle lying next to McCracken's body. "[I] had to wipe the blood off it," Stevens recalled. Once the rest of the battalion crossed, Hamilton sent a company south in search of COL Barth.

They would not have trouble finding him. Earlier, Barth had led his two battalions down to the river's edge encountering only sporadic rifle fire but finding the two bridges destroyed. When he saw truck-loads of Germans fleeing the city he decided to immediately cross the river without waiting for rubber assault craft to reach his position. His men found a skiff and a large leaky boat for their amphibious assault. They improvised oars by tearing down a nearby fence and using its boards. The men were unsure of their fleet's seaworthiness but Barth reassured them by crossing with the first flotilla. Once the boats had crossed, Barth and an engineer paddled the boats back to the west bank to manage the operation. As they reached the eastern bank, the men climbed out of the boats and up the wooded hillside.

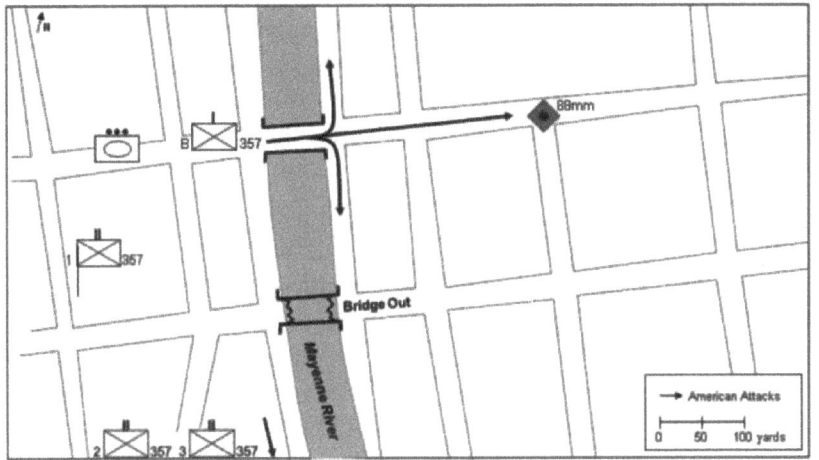

TF Weaver seizes the bridge at Mayenne.

One of the men, CPT Max Kocour, spotted a French farmer who put his finger to his mouth to make the "quiet" sign and pointed to a certain farmhouse door. Kocour cocked both his M3 "grease gun" and his .45-cal. pistol before quietly creeping along the wall to the door. Peering inside, he spied three Germans sitting around a table eating lunch with their rifles stacked near the door.

Kocour quickly stepped in the doorway and placed himself between the Germans and their weapons. "Hande hoch!" he ordered. The Germans only half-heartedly obeyed. A now furious Kocour shouted, "Patton, ser gross panzer! Erschiesen sie! (Patton has large tanks. They will shoot you!)" It worked. The Germans threw their hands up. The Frenchman and four of his friends collected the rifles and helped Kocour escort the prisoners down to the river. Kocour sang the *Star Spangled Banner* as he walked while the Frenchmen cheered "Vive Americain!"

While Kocour was capturing his quarry, inflatable rafts finally arrived on the west bank and more men crossed. By 1030, two battalions were safely across and making their way east to block the roads leading into Mayenne. In crossing the river in the makeshift vessels, COL Barth had made a risky decision that paid huge dividends. "We took a chance and were lucky," remembered Barth.

Back at the bridge, 1LT Stevens reorganized his scrambled units and directed them to protect the bridgehead. 1LT Lombardi, who had led the attack in his lone tank, went back to replenish his ammunition. More men, tanks, and tank destroyers crossed the bridge and spread out, expanding the bridgehead. They reached some railroad tracks where they observed a German tank but before the men could engage, the tank turned and rushed out of the city. Stevens' men pushed through the town until they made contact with Barth's men. The linkup was complete. Mayenne and its bridge were safely in American hands. The Americans immediately established a defensive perimeter east of the city.

As it grew dark, BG Weaver ordered a halt and organized a defense of the city. All night long, Germans stumbled into 90th Division positions. The Americans captured large numbers of bewildered Germans who could not believe their enemy was so far forward. When four German vehicles drove into La Ferichard Hotel, which served as COL Barth's command post, an American platoon knocked out each vehicle and rounded up the prisoners right outside Barth's front door. "The Germans apparently had no idea where we were and stray vehicles kept barging into the town," explained Barth.

Hamilton also found himself face-to-face with confused Germans. Just after 2300, two German vehicles rolled up to his command post and three men got out of the first car. They were immediately fired on and the Germans surrendered. "Suddenly," recalled Hamilton, "we were fired on from behind the two vehicles." The two German officers from the second vehicle had been ducking behind the first, hoping to escape. At that moment, a jeep mounting a .50-cal. machine gun shot around a corner and fired on the second vehicle, setting it on fire. Two captains, wielding Tommy guns, then ran around one side of the vehicles, killing and wounding all the Germans. The fire Hamilton encountered had been from his own troops. Their work had been accurate and intense. "The dead officer was so full of holes that he grotesquely resembled a sieve," explained Hamilton.

The tally of German vehicles destroyed included seven cars, two motorcycles, and one truck. The Americans also destroyed two 88 mm artillery pieces. The Germans were dazed and confused. Well west of the town, German bombers dropped flares on suspected Allied positions, unaware how far east the Americans had advanced.

The next morning, 6 August, the citizens of Mayenne emerged from their hiding places. One group unfurled a special flag. It was a homemade

Stars and Stripes the locals had sewn to welcome American soldiers during World War I. They had hidden the flag for the last four years until they were sure the Germans were gone. The flag represented the town's official liberation.

While the victory at Mayenne was not perfect, officers and NCOs proved capable of inspiring their men to achieve the unit's goals. The bridge over the Mayenne River proved vital for the Allied drive surrounding the Germans in the Falaise Pocket, an operation which resulted in the killing or capture of close to 50,000 German soldiers and the destruction of over 300 tanks and other armored vehicles. The 90th ID was one of the few units that actually remained in Third Army for the entire European campaign. At the end of the war Patton claimed that the division was among his very best.

For Further Reading

John Colbey and Melissa Robert. *War From the Ground Up: The 90th Infantry Division in World War II,* Eakin Pr., 1991.

John C. McManus. *Americans in Normandy: The Summer of 1944--The American War from the Normandy Beaches to Falaise*, Forge Books, 2005.

Martin Blumenson. *Breakout and Pursuit: US Army in World War II: The European Theater of Operations*, Office of the Chief of Military History, Department of the Army, 1961.

One of the destroyed bridges in Mayenne, south of the bridge seized by the 1st Battalion, 357th Infantry.

The National Archives
SC 192484 (Signal Corps)

The Six Principles of Mission Command

1. Build Cohesive Teams through Mutual Trust
2. Create Shared Understanding
3. Provide a Clear Commander's Intent
4. Exercise Disciplined Initiative
5. Use Mission Orders
6. Accept Prudent Risk

Principles of Mission Command Illustrated in the Mayenne case

1. Build Cohesive Teams through Mutual Trust. Since landing in France in June 1944, the 90th Division had experienced tough combat and had seen its morale drop. LTG Patton chose to change the leadership of the division by appointing BG McLain as the commander and assigning BG Weaver as assistant division commander. Both men had an excellent reputation as tough warriors, which served to start rebuilding trust among subordinate leaders in the division. McLain reinforced the need for command competence by relieving 16 ineffectual field officers in the first several days of his tenure. Because his immediate mission required aggressive offensive operations against the Germans, he built a task force with an aggressive commander, BG William Weaver, at the helm. That decision served to enhance cohesion among the units of the 90th Division.

2. Create Shared Understanding. When BG McClain called COL Barth, the 357th Infantry Regiment commander, to his headquarters, he ensured that the larger goal of the upcoming Mayenne operation was clear. Barth's regiment – and the other elements of the task force – would need to seize the city's bridges so that US forces could continue their drive through German forces.

3. Provide a Clear Commander's Intent. BG McClain articulated his intent clearly when he provided COL Barth a simple mission: capture Mayenne and seize its bridges intact. The details of how this objective was to be attained were left to Barth and BG Weaver to figure out.

4. Exercise Disciplined Initiative. The operation to seize the bridges at Mayenne offers numerous examples of disciplined initiative. Early in the operation, MAJ Hamilton altered the plan to cross the bridge by halting the artillery barrage before it was complete in order to use the German smoke to conceal the American crossing. Seeing the lead platoon balk at crossing the bridge, 1LT Stevens showed great personal courage and led the way across. He later pushed men out of their hiding places

and directed the platoon sergeant to get his unit to the eastern side of the river. At a higher level, while the intact bridge was under assault, Colonel Barth immediately tried to find ways across the river after discovering that the two southern bridges had been destroyed. Barth immediately secured skiffs and other watercraft to begin ferrying his men across the river, rather than wait for assault boats. This tedious action assisted in expanding and securing the bridgehead on the eastern side the river.

5. Use Mission Orders. BG McLain did not tell his subordinates how to capture the bridges at Mayenne, only that they had to take them. In addition, BG Weaver let MAJ Hamilton develop his own plan of action for capturing the bridge intact, who in turn, came up with a sound attack plan that despite some difficulties, succeeded. BG Weaver also used mission orders when he ordered COL Barth to envelop the town from the south. He did not tell Barth how to do it, but merely gave him general guidance on what was to be achieved.

6. Accept Prudent Risk. The bridges over the Mayenne River had to be taken intact. BG Weaver accepted reasonable risk by dividing his forces and planning a two-pronged attack. This diluted his combat power but by sending COL Barth with two battalions to the southern side of the town to envelop enemy forces, he hoped to confuse the Germans and perhaps unhinge their defenses. At the intact bridge, MAJ Hamilton accepted risk in terminating the artillery barrage early in order to mask his attack with the smoke that obscured the bridge.

The Victory at Tarin Kowt

Donald P. Wright, Ph.D.

The Coalition campaign to overthrow the *Taliban* regime in Afghanistan began in late 2001, just a month after the 9-11 terror attacks on New York City and the Pentagon. Because of Afghanistan's remoteness, US commanders chose to partner with anti-*Taliban* militia leaders within the country itself as the main weapon of choice to overthrow Taliban power. To enable these militias, the US inserted a number of Army Special Forces teams that would advise the militia commanders and coordinate Coalition airpower in support of operations against *Taliban* forces.

The teams sent into Afghanistan, formally known as US Army Special Forces Operational Detachment-Alphas (ODAs), were trained specifically to work with foreign military forces. Each ODA was made up of 12 Soldiers, two officers and 10 non-commissioned officers. Each Soldier had a particular role to play in team operations and was highly trained in his occupational skill. To qualify for Special Forces, a Soldier not only had to be physically fit and competent with his weapon and equipment, he also had to be highly intelligent and adaptive. ODAs were expected to live with foreign soldiers, adapt to their cultural norms, and form close bonds with them. Often operating in isolation from other US forces, Special Forces Soldiers had to be able to make sound decisions quickly based on information they collected from their partners. The ODAs sent into Afghanistan in the fall of 2001 were expected to do all of these things. Most importantly, they had broad latitude to work with Afghan militia leaders to obtain the Coalition's ultimate objective of the end of *Taliban* power.

On 14 November 2001, ODA 574 along with personnel from the Air Force and other US organizations arrived in southern Afghanistan to link up with the Afghan political leader Hamid Karzai and his small anti-*Taliban* militia. Although of Pashtun ethnicity, like the great majority of the Taliban, Karzai hoped to lead the fight against the *Taliban* in southern Afghanistan. The team's mission was to "infiltrate the Oruzgan province, link up with Hamid Karzai and his *Pashtun* fighters, and advise and assist his forces in order to destabilize and eliminate the *Taliban* regime there."

On his arrival, the ODA leader CPT Jason Amerine sat down with Karzai to establish a relationship with him and understand the situation as Karzai comprehended it. Amerine had briefly met Karzai in Pakistan one month earlier when he accompanied a more senior US commander in initial discussions on how they could work with the *Pashtun* leader in the fight against the *Taliban*. Now that they were both in Afghanistan, Karzai told the American officer that the key to winning Kandahar as well as Uruzgan province was to capture the town of Tarin Kowt located 75 miles north of Kandahar city. Hamid Karzai described Tarin Kowt as the heart of the *Taliban* movement and all the major leaders of the *Taliban* movement had families in and around Tarin Kowt. Mullah Omar was from Deh Rawod which was just to the west of Tarin Kowt. The seizure of Tarin Kowt would therefore strike a blow to *Taliban* morale. Further, Karzai believed that by taking Tarin Kowt, all of the *Pashtun* villagers would essentially surrender or turn completely to his cause.

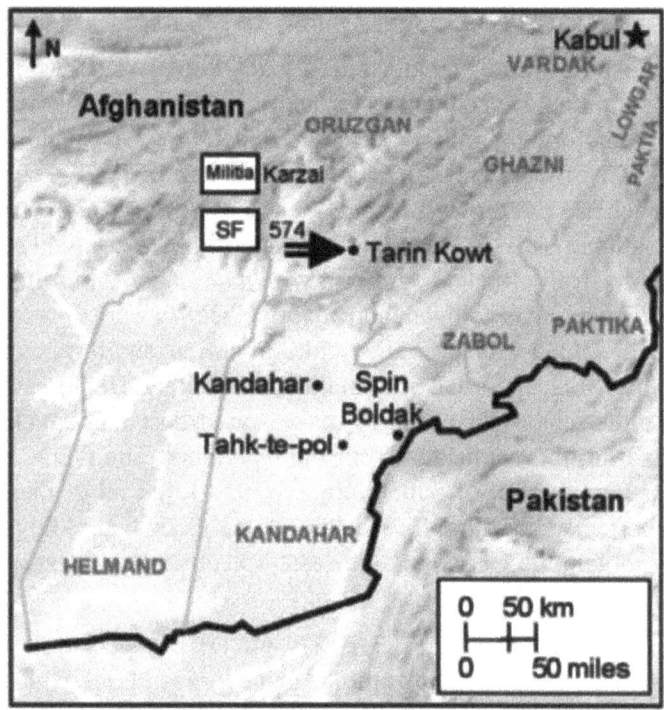

Figure 1. Southern Afghanistan and ODA 574/Karzai.

Amerine then gathered his team, pulled out some maps, and developed a plan to take Tarin Kowt. That plan amounted to a siege. Karzai's forces along with their SF advisors would close off the mountain passes leading into the town. Karzai had reasoned that once that was accomplished, the town would simply surrender. Additionally, he informed CPT Amerine that there were already friendly fighters in Tarin Kowt who would foment an uprising if necessary. Given the small numbers of troops that were available, his ODA and the 150 Afghan fighters in Karzai's small group, Amerine told Karzai that they would have to create a larger force but this would take time and Amerine imagined a long period of arming and training Afghan volunteers before the town could be taken. While halted at a small village some 30 miles southwest of Tarin Kowt, ODA 574 began arranging for the shipment of more weapons and ammunition. Hundreds of people arrived to try and get weapons but most were only interested in protecting their own homes and villages. With the recruiting effort just starting, news arrived on 16 November that stunned Amerine and his Soldiers. Key leaders in Tarin Kowt had risen up against the *Taliban* governor, killed him and his bodyguards, and seized control of the town. They wanted help from Karzai and the Americans especially because they expected the Taliban to act quickly and forcefully restore their control of the town.

The news of the uprising presented Amerine with a dilemma. If they moved into Tarin Kowt and the Taliban launched a counterattack, Karzai's forces were too small to defend the town. It was doubtful that in the short time they would have before an attack, enough reliable and capable volunteers could be recruited. Still, Amerine knew he had the trump card of American air power on his side. It was a difficult choice but Amerine and ODA 574 decided to support Karzai's insistence that they go immediately to Tarin Kowt and take advantage of the military and political opportunity.

Piling into a motley collection of beat up trucks and other vehicles sent by village elders, the ODA and their Afghan partners bounced along the mountain roads to the town. En route, Karzai worried that the population of Tarin Kowt might be angry that American Soldiers had accompanied his force to the town. His fears were quickly quelled though when the town's people warmly welcomed the Soldiers.

Once in the town, Karzai met immediately with the town leaders who had led the uprising, leaving military matters to ODA 574. He stayed busy getting in touch with other *Pashtun* leaders in the area, constantly recruiting fighters and supporters, and undermining the *Taliban*'s rule in the process. Many of the area's leaders came to speak with Karzai and offer information on *al-Qaeda* and *Taliban* elements near the town. He also discovered that many of the Muslim clerics in the region were supportive of his actions. Early that evening, other informants brought him the news that he had been expecting, a large force of *Taliban* was en route to Tarin Kowt.

Karzai quickly requested that Amerine meet him and his local supporters so that they could explain the situation. The Afghan leaders proceeded to matter-of-factly mention that hundreds of Taliban troops were approaching the town and that the enemy force, mounted on a large number of trucks, would probably arrive sometime the next day. Amerine remembered, "It took me a second to digest it. At that point, I said, 'Well, it was nice meeting all of you. I think we need to organize a force now and do what we can to defend this town.'" The Special Forces captain attempted to excuse himself so that he could start preparing to oppose the threat. His Afghan hosts, however, would not hear of it. Since it was the first day of *Ramadan*, they insisted that he stay, drink tea, eat, and talk. Sensing that he could not embarrass his hosts, Amerine stayed just long enough to satisfy their request, then quickly made his exit but not before asking Karzai to send every fighter he could find to the ODA's headquarters as soon as possible.

Returning to his men, Amerine pulled them together and told them about the impending arrival of the *Taliban* forces stating, "Well they're coming from Kandahar. We know it's a large convoy." The captain then ordered a number of actions. His communications sergeant began contacting the team's headquarters to inform them about the imminent assault. The team's Air Force enlisted terminal attack controller (ETAC) passed warning orders through those channels to let the Air Force and Navy know that their support would soon be required at Tarin Kowt. Amerine's team worked into the night to arm all the new Afghan fighters that showed up and develop a plan to hold the town.

After midnight on 17 November, after receiving a report that Coalition aircraft had spotted a convoy moving north toward Tarin Kowt, Amerine ordered the aircraft to drop bombs on what he believed was the lead element of the Taliban attack. He then took his ODA and 30 Afghan fighters out of the town on the main road leading south. He spotted a plateau approximately eight miles from Tarin Kowt from which the team could observe one of the roads from Kandahar as it came through a pass at the south end of a broad valley. Amerine established what he called the Overlook Observation Point on the plateau. The *Taliban* convoy traveling to Tarin Kowt would have to enter the valley from the pass on the other side. Excellent fields of observation across the valley would allow the ODA and Karzai's men the opportunity to use close air support and other weapons to engage the *Taliban* before they even reached the outskirts of Tarin Kowt.

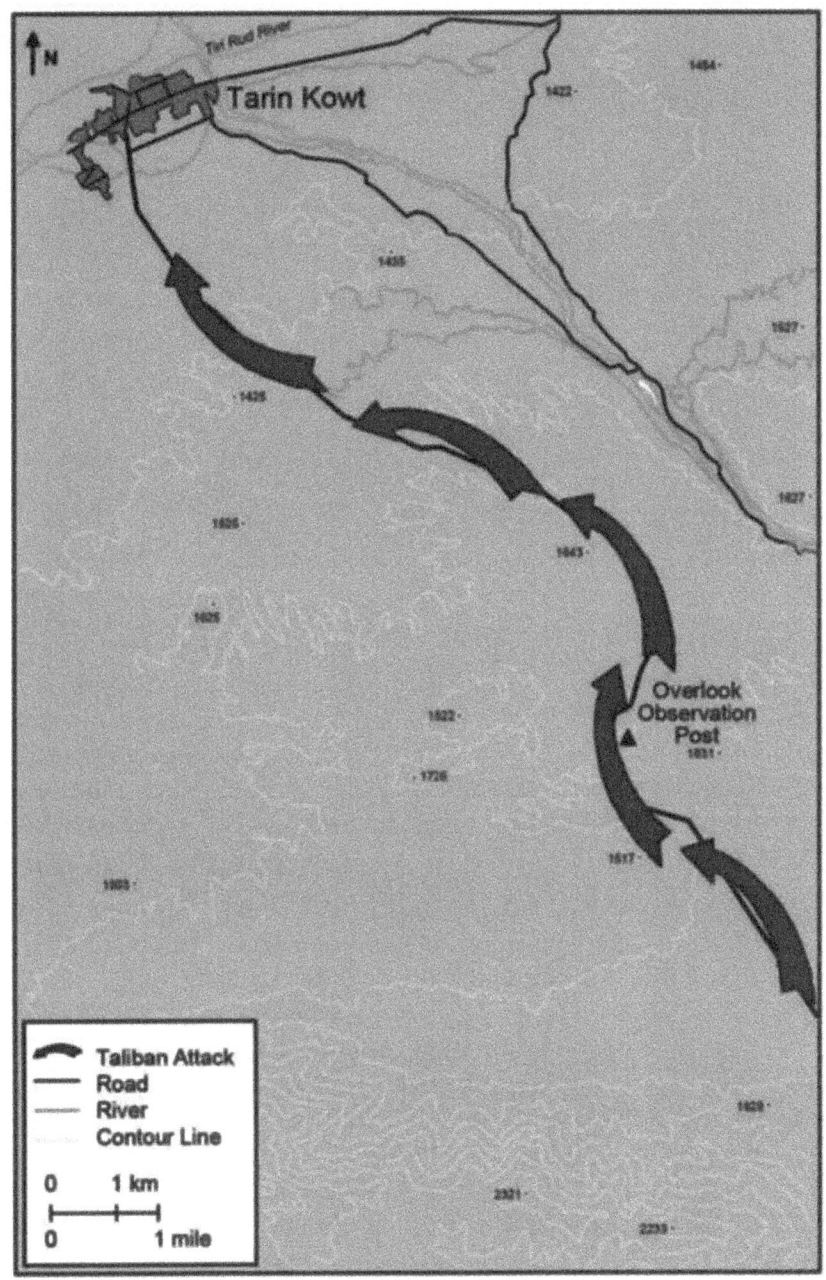

Figure 2. Tarin Kowt/Overlook Observation Point.

ODA 574 with Hamid Karzai. CPT Amerine in back row, second from the right.
US Army photo.

Early that morning, Amerine and his team suddenly saw dozens of *Taliban* vehicles emerge from the pass and spread out on the valley floor heading north. Amerine told his combat controller to begin bringing in the close air support. Using a laser designator, the team's ETAC directed the first bomb onto the *Taliban* formation destroying a vehicle in the lead element. It began to look like ODA 574 was about to defeat the *Taliban*.

Then an inexplicable event occurred, one that Amerine described as feeling like "we were seizing defeat from the jaws of victory." Karzai's men panicked. The lack of training among these militiamen demonstrated itself with graphic clarity when they perceived that the battle was not going well and their best option at that point was to withdraw to Tarin Kowt. To make matters worse, without Karzai or any other English speakers at the observation point, the men of ODA 574 could not communicate with the panic-stricken Afghan tribesmen. The Afghans hopped into the vehicles and were only prevented from driving off immediately by the members of ODA 574 who literally stepped in front of the vehicles to get them to stop. If the trucks left, the Americans had no way to get back to Tarin Kowt. Reluctantly, the Soldiers of ODA 574 jumped aboard the trucks and went with their charges back to the town.

With the *Taliban* convoy continuing its advance toward the town, ODA 574 and Karzai had to turn the situation around. In Tarin Kowt, the team met with Karzai and after a quick consultation, ODA 574 sped south of town again to establish another observation post which they could use for calling in close air support. They found a new site much closer to the town and once again, Amerine's team began calling in airstrikes on the approaching enemy forces. At the same time, Afghan fighters began appearing near the ODA's position to help defend the town.

With the renewal of the attacks on the *Taliban*, ODA 574 ran into a new and wholly unexpected problem, crowds of civilians from Tarin Kowt had begun arriving to watch the battle. The ODA had not expected to have to deal with this type of situation. CPT Amerine called it a "circus atmosphere" in which Afghan children attempted to rummage through their equipment and older civilians meandered around the defensive position. One member of ODA 574 pleaded with an English-speaking Afghan to at least send the children back to Tarin Kowt because of the danger of the situation. Thankfully none of the townspeople was injured as the pace of the attacks on the *Taliban* convoy increased.

Initially, the lead trucks were targeted to slow the convoy down. When those vehicles were destroyed, the Coalition aircraft simply began working their way back through the convoy which was now very spread out. When they ran out of bombs, several pilots began making strafing runs against the *Taliban* vehicles. While the enemy columns had approached their position, the men of ODA 574 were relieved to see that the aircraft had ultimately stopped the attack's forward momentum.

Sometime after 0800, another unexpected surprise struck the ODA. Two of the *Taliban* trucks had found an alternate route into Tarin Kowt. The American troops began to hear small arms fire on their flank which indicated the enemy was close by. The mounting gunfire caused Amerine to think that perhaps the battle was not yet won. Unbeknownst to him, a number of villagers had moved to the threatened area and fought off the *Taliban* attackers. That action actually signaled the end of the battle. For the next two hours, the remnants of the convoy took hit after hit from close air support (CAS) sorties as the *Taliban* tried to make their way back to Kandahar.

Before and during the battle, Karzai had become concerned about the perception among his fellow Afghans that the US Soldiers had precipitated the *Taliban* attack on Tarin Kowt. On the day of the attack, one of the local mullahs paid a visit to Karzai to speak with him. Karzai was anxious that the mullah, who spoke for others, was going to tell him that the *Taliban* attacked the town because there were Americans in Tarin Kowt and that

Karzai and his supporters must leave. If this statement was communicated, Karzai was certain that others in the region would turn against his efforts to liberate southern Afghanistan from the *Taliban*. His fears were thankfully dashed when the mullah instead told him, "If the Americans hadn't been here, we would have all been killed." That statement was an indication that the military victory had also become a political success.

ODA 574 and Hamid Karzai's small force, assisted greatly by Coalition air support, had clearly triumphed over the *Taliban* at Tarin Kowt. COL John Mulholland, the senior US Special Operations officer in Afghanistan, later viewed the engagement at Tarin Kowt as "pivotal" to the Coalition's efforts in southern Afghanistan. Furthermore, Mulholland argued that the *Taliban* recognized the potential threat posed by Karzai to their legitimacy in the region and tried to destroy Karzai and his supporters. According to Mulholland, when that attack failed, the *Taliban* grew greatly concerned about their hold on the south.

This belief seemed borne out by the success Hamid Karzai enjoyed in rallying other Pashtuns to his cause. CPT Amerine not only witnessed firsthand the destruction of the *Taliban* forces, he also saw the reaction of other *Pashtun* Afghans to Karzai. He realized the tremendous psychological and political importance the victory had and its resulting impact on the enemy. Karzai's tireless work in securing political support from the various groups in the Tarin Kowt area, and elsewhere as it would turn out, made ODA 574's future tasks less difficult. Amerine explained that this support translated into rapport and trust with the *Pashtuns* in the area and enabled him and his team to look ahead to the next task of the liberation of the city of Kandahar.

The battle of Tarin Kowt was clearly an instance where a well-trained and mentally agile group of Special Forces Soldiers combined with air power, indigenous partners, and a deep understanding of commander's intent, achieved a major victory. This was a textbook example of how a small well-trained force could employ unconventional warfare for a superlative result.

For Further Reading

Eric Blehm. *The Only Thing Worth Dying For: How Eleven Green Berets Fought for A New Afghanistan* (2010)

Donald P. Wright, et al. *A Different Kind of War: The US Army in Operation Enduring Freedom, October 2001 – September 2005* (2010)

Charles H. Briscoe et al. *Weapon of Choice: ARSOF in Afghanistan* (2003)

The Six Principles of Mission Command

1. Build Cohesive Teams through Mutual Trust
2. Create Shared Understanding
3. Provide a Clear Commander's Intent
4. Exercise Disciplined Initiative
5. Use Mission Orders
6. Accept Prudent Risk

Mission Command in the Tarin Kowt case

1. Build Cohesive Teams through Mutual Trust. ODA 574 was a small unit composed of very experienced and well-trained Soldiers. The team had been training in Kazakhstan when 9-11 occurred. They then returned to FT Campbell and prepared specifically to go into Afghanistan and work with anti-Taliban forces. Each member knew his team mates well and trusted them implicitly.

2. Create Shared Understanding. ODA 574's higher headquarters had established a clear vision for the endstate in southern Afghanistan. It focused on winning over Pashtun support and ultimately taking control of Uruzgan province and Kandahar city. CPT Amerine understood this and had his vision of the mission reinforced when he briefly met with Hamid Karzai in Pakistan in October 2001 before both men moved into Afghanistan. It was clear to Amerine what he had to do and even who his Afghan partner would be. Throughout their time in southern Afghanistan, Amerine and Karzai consulted and reassured one another of their mission and shared goal.

3. Provide a Clear Commander's Intent. The team's mission was to "infiltrate the Uruzgan province, link up with Hamid Karzai and his Pashtun fighters, and advise and assist his forces in order to destabilize and eliminate the Taliban regime there." This mission was articulated so broadly that it resembled a commander's intent, in that it emphasized the "what" rather than the "how" and "when." As such, it left a great deal of latitude for freedom of action. CPT Amerine and his team would take advantage of this latitude as the political and military situation changed radically and rapidly around them.

4. Exercise Disciplined Initiative. The success in seizing and securing Tarin Kowt was made possible by ODA 574's exercise of disciplined initiative. The two best examples of this was CPT Amerine's reliance on Karzai's political advice about need to take Tarin Kowt before moving

toward Kandahar, and ODA 574's immediate move into Tarin Kowt despite concerns about Taliban counterattack against the small US/Karzai force. Amerine did not seek permission for these actions from his higher headquarters. Instead, he acted within his understanding of his commander's intent and the trust he had in his Afghan partner, Karzai.

5. Use Mission Orders. As noted above, ODA 574's mission statement was "to infiltrate the Uruzgan province, link up with Hamid Karzai and his Pashtun fighters, and advise and assist his forces in order to destabilize and eliminate the Taliban regime there." This statement emphasized the "what" and the endstate rather than the "how" and "when." With the political landscape of southern Afghanistan so much in flux, CPT Amerine was given a broad mission statement that allowed him great latitude in making decisions based on the conditions he found on the ground. This enabled the great success he and his team had in Tarin Kowt.

6. Accept Prudent Risk. Without question, the riskiest decision CPT Amerine made in this action was to move into Tarin Kowt with only his ODA and a small band of Afghan militia to face a Taliban counterattack. The risk was mitigated by his ability to direct Coalition airpower against the Taliban. That close air support in fact defeated the Taliban attempt to retake Tarin Kowt, ensuring that the Coalition effort in southern Afghanistan would continue. The decision also helped build rapport with Karzai who was sure it was the right move based on his understanding of the political situation. In the fall of 2001, the combination of Coalition firepower and Afghan political leadership proved to be a winning partnership in southern Afghanistan, as it had been in the north of that country.

The Attack on the Ranch House, August 2007

John J. McGrath

On 22 August 2007, insurgents conducted a deliberate attack against a combined US Army-Afghan combat outpost (COP) outside the village of Aranas, called COP Ranch House, in central Nuristan Province. The attack against half a platoon of troops from C Company, 2-503d IN, 173d Airborne Brigade, was initially successful with the attackers breaking through the outpost's perimeter in a section manned by paramilitary local Afghan Security Guards (ASG). However, the defenders fought back vigorously and soon repulsed the attack with no American fatalities. Key insurgent leader Hazrat Omar was killed in the attack. At the time the Ranch House position was perhaps the most remote American outpost in Afghanistan, located in the rugged southern foothills of the Hindu Kush range in an area isolated from roads and rivers. Because of this remoteness, the American command had planned to close the outpost before the attack and it was, in fact, closed in October 2007. After its closure, many of the American troops involved in the action were ambushed southwest of the former post while returning from a foot patrol to Aranas on 7 November 2007. In this action, six Americans were killed.

Aranas is the largest community in the central Nuristani district of Waygal. The town sits on the south-facing northern slope of an eastward running valley that branches off from the Waygal River, the major terrain feature in the district, several miles to the south east. The Waygal flows south from the Hindu Kush into Kunar Province joining with the Pech River at Nangalam about ten miles southwest of Aranas. In 2007, Aranas and its outlying area had about 6,000 inhabitants making it a metropolis for the area. The population of Nuristan is a unique ethnic group, neither Iranian (like the *Pashtuns* who lived to the south in Kunar) nor *Indic* (like the people who lived to the east in Pakistan) but distantly related to both groups. Nuristan is an isolated area even for Afghanistan. Governmental control had long been a weak and distant concept to the Waygali Nuristanis.

Aranas had long been recognized as a hotbed of insurgency by American intelligence analysts, providing an area from which the enemy could launch attacks into the vital Kunar and Pech valleys from long established base camp areas. Therefore, as part of a long term counterinsurgency strategy, the Coalition leadership felt it was essential to establish an American-Afghan outpost in the area, both to limit insurgent activities in the region and to use as a base from which to conduct counterinsurgency activities among the local population.

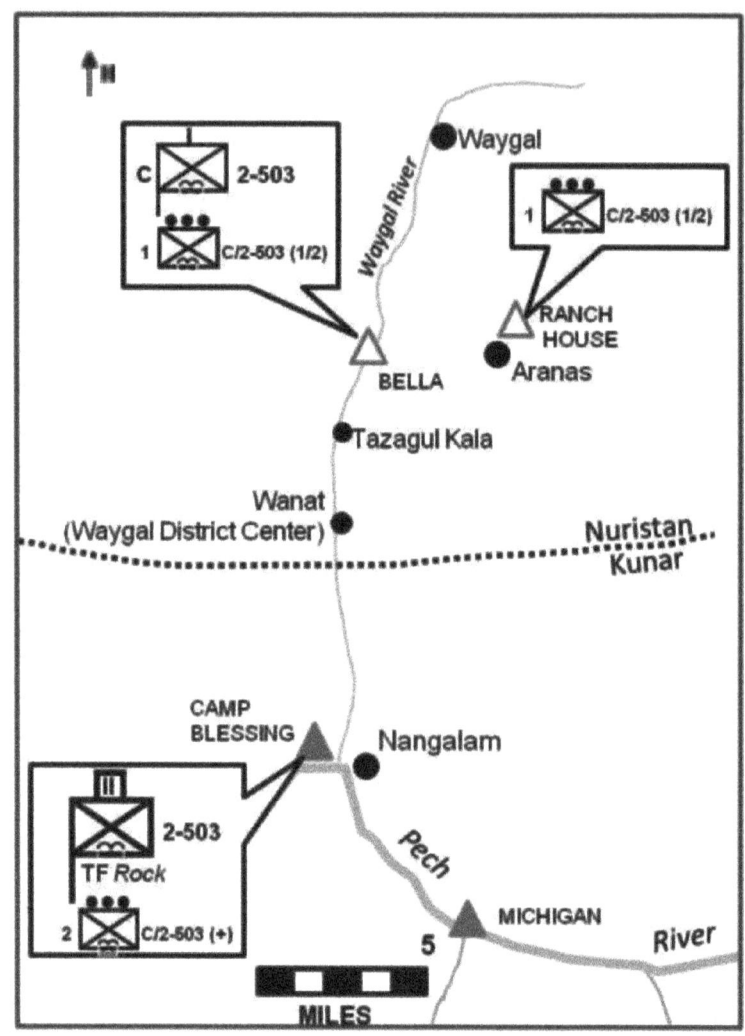

Figure 1. Coalition positions in the Waygal Valley, 2007.

American operations in the area were extremely limited before 2006. A forward operating base (FOB) called Camp Blessing was built near Manoguy at the point where the Waygal flowed into the Pech River in 2003 about 15 miles south of Aranas in Kunar Province. A pair of 105-mm howitzers and several 120-mm mortars were placed at the outpost, which was initially manned only by a small Special Forces element and later a platoon of US Marines. The 105-mm guns were later replaced by two 155-mm howitzers. The situation changed in 2006 when troops from the 3d Brigade, 10th Mountain Division (Task Force *Spartan*), and the

3d Brigade, 201st Corps of the Afghan National Army, deployed into a number of newly created or expanded FOBs (forward operating bases) and combat outposts in Kunar and Nuristan Provinces.

Afghan Soldiers in front of The Ranch House.
US Army photo.

TF *Spartan*'s 1st Battalion, 32d Infantry (1-32 IN), established two outposts in the Waygal Valley in August 2006. One of these, COP Ranch House, was located just northeast of Aranas on a high 7,000 foot mountainside. The position was centered on a large one-story wooden building, a former schoolhouse that resembled the Ponderosa ranch house on the 1960s television show *Bonanza*, a comparison which gave the outpost its name. The site was virtually impossible to reach by motorized vehicle and the helicopter landing zone (LZ) was placed on top of the Ranch House building, the only available area flat enough to accommodate an aircraft. Since the building backed up against the steep mountainside, the Americans had to use explosives and engineer equipment brought in by sling load to carve out an adequate LZ. Eventually an LZ large enough to accommodate a UH-60 Blackhawk helicopter, the aircraft used for medical evacuations (MEDEVAC), was created. The larger CH-47 Chinooks, the mainstay for resupply missions, still had to sling-load cargo above the LZ in order to deliver its loads. With no roads suitable for HMMWVs, the unit often used donkeys for ground resupply as well. In addition to the Ranch

House, 1-32 IN established a second position at a small hamlet known as Bella on the Waygal River southwest of the Ranch House, roughly a third of the distance between Aranas and Camp Blessing or three kilometers (1.5 miles) as the crow flies but seven kilometers (four miles) by foot trail, southwest of the Ranch House. The American units split a platoon between both outposts in 2006 and 2007, placing approximately 20 troops at each position.

At the Ranch House outpost, the 1-32 IN troops built a defensive perimeter to the northeast of the Ranch House building eventually consisting of a series of six positions, generally of sandbagged wooden towers and concertina wire which encircled the whole post. Americans manned four of the positions which were numbered 1 to 4 starting from the north extending to the east and around back to the west. A fifth position was later added between posts 3 and 4 to the southeast of the perimeter. This post was manned by members of the Afghan Security Guards (ASG), a locally recruited force. A small detachment of Afghan soldiers (Afghan National Army or ANA) manned a guard post built onto the north side of the Ranch House building. Two additional ANA positions were located directly behind and to the left and right of the ASG site (Post 5) between it and Post 4.

In the center of the position was a small aid station, a bunker used as a tactical operations center (TOC), a rations supply point and a mortar firing position equipped with a 60-mm mortar from the company mortar section. The ASG also established an observation post two kilometers northwest of the Ranch House on a mountaintop that was visible from both the Ranch House and Bella and which was used to provide overwatch for patrols travelling between the two outposts.

Although located only 20 kilometers (12 miles) northeast of Camp Blessing, the rugged terrain around the Ranch House made the position arguably the most remote in Afghanistan. The site ran along an east-west running spur with the eastern end higher than the western portion. The outpost was built on a slope that rose about 300 meters in elevation from the low point at the LZ to the highest position, Post 3. The elevation at the Ranch House building was about 7,300 feet. The slope continued beyond the end of the outpost perimeter several hundred meters to a ridgeline that was at an elevation of 8,400 feet. The position was located within 25 meters of the outlying houses of Aranas proper to the southwest and had several cottages or *bandeh*s located on the slopes surrounding the other sides of the perimeter.

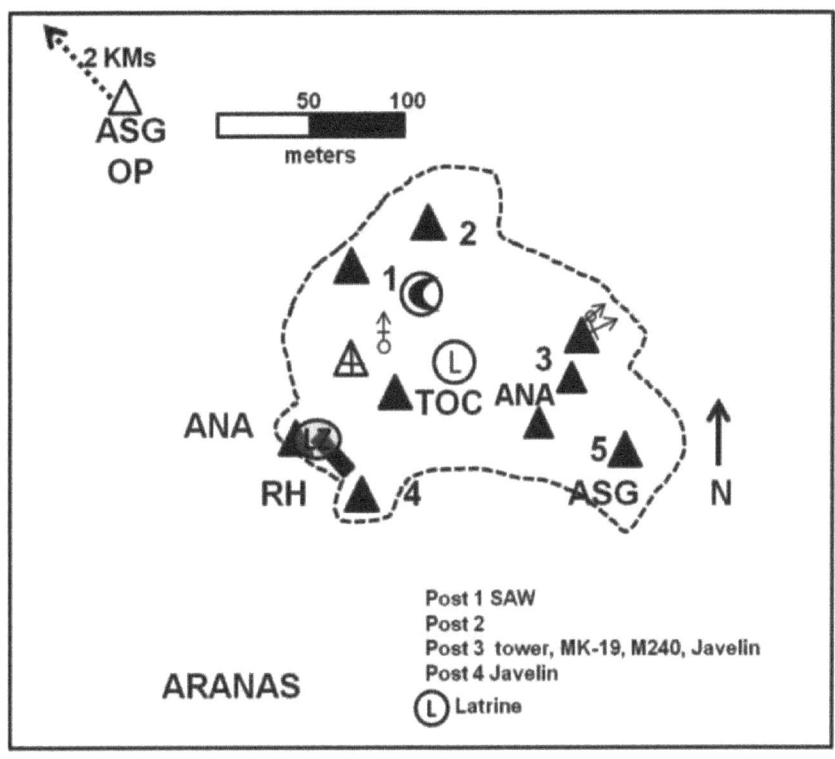

Figure 2. The Set Up of COP Ranch House.

The operations at the Ranch House outpost were part of the ongoing counterinsurgency campaign. The troops at the COP conducted a mix of missions including securing the local population, providing humanitarian assistance, and establishing a presence of the central Afghan government. As such, the commander's intent was to ensure that the outpost could sustain and protect itself in order to conduct such tasks. The combat outpost needed to be adequately defended both to deter insurgent activities in the area and to instill confidence among the inhabitants to support the activities of the Coalition and the Afghan government.

When Task Force *Rock*, the 2d Battalion, 503d Infantry (2-503 IN), 173d Airborne Brigade, took over the Ranch House outpost on 26 May 2007, the position had not experienced any direct enemy contact since March. The new unit modified the defenses slightly by emplacing more Claymore mines around the perimeter than their predecessors had, shifting the perimeter concertina barbed wire farther out in several locations, and adding more sandbags to the positions and building alternate fighting positions in case any of the existing positions were destroyed in a large enemy attack. Because of the difficulty in resupplying the position, the perimeter wire was only a single strand of concertina fence which was stretched around obstructions such as large rocks.

Shortly after the 2-503 IN took over the Ranch House in May 2007, the outpost was fired upon while some of the defenders were conducting a nighttime ambush patrol. The enemy directed PKM machine gun and RPG rocket fires at the ASG position, Post 5, for about an hour. This post was considered to be a weak link in the defense as a steep slope overlooked the position from the east. After this encounter, the outpost remained quiet for almost three months while other areas in the TF Rock sector, particularly the Korengal and Pech valleys, had almost daily contact with the insurgents. Meanwhile, the Ranch House garrison frequently heard rumors of impending attacks, none of which materialized. The lack of action and the repetitive false warnings meant that when an attack did come, the defenders, while prepared, were nevertheless surprised.

The ASG contingent in the Waygal Valley had been established by the 1-32 IN. This force was locally recruited and given only rudimentary military training. ASG members were used as scouts and to buttress defensive positions by manning some positions particularly observation posts. Just before its departure, the 1-32 IN had expanded the size of the ASG contingent from 25 to 45. The recruitment of the ASG was done to give the local population an investment in the Coalition presence in the area as well as provide an economic boost. However, the ASG troops and leadership soon proved to be unreliable.

As noted earlier, the five numbered posts included four manned by Americans. The posts were built out of plywood lumber and sandbags and, in most cases, were combination guard posts and sleeping quarters. The guard posts were built up on elevated towers equipped with crew served weapons. The weapons were a combination of Mk-19 automatic grenade launchers, M240B machine guns, Squad automatic weapons (SAW) and Javelin antiarmor missile systems. Posts 3 and 4 had sleeping quarters under the tower which had no direct entrance to the tower from the sleeping area. Posts 1 and 2 had offset sleeping areas. While the sleeping quarters for Post 1 were only 15 meters away, Post 2's was a relatively distant 50 meters. In addition to reinforcing positions, sandbags were used to build staircases on the steep hillside to facilitate troop movement. Unlike at other sites in northeastern Afghanistan, the perimeter of the Ranch House combat outpost was not made of HESCO fabric barriers. The prefabricated

HESCOs were bulky and required extensive dirt fill and a Bobcat front end loader to fill them. While there was a Bobcat at the site, the terrain did not contain adequate amounts of dirt and the slope restricted the effective use of the Bobcat to the landing zone area. Plywood and sandbags provided the fortifications at the Ranch House.

Despite its location, the Ranch House outpost had excellent communications systems on site which in the summer of 2007 including a tactical satellite radio (TACSAT) and secure internet protocol router (SIPR)-capable very small aperture terminal (VSAT). The VSAT provided classified messaging. The SIPR and non-secure communications, which depended on satellite connections, were sometimes down due to weather conditions. On one occasion these communications means were briefly lost when an RPG round hit the antennas. At the time of the attack, the communications were functional. An unnamed US Army signals specialist, properly referred to as a SIPR point of presence (SPOP) technician, was at the outpost during the attack. The TOC also had FM radio communications with each post and with company commander CPT Matthew Myer at Bella. An antenna array, including a large satellite dish, sat on top of the TOC bunker.

An array of fire support assets was available to the Ranch House defenders. The outpost itself contained one 60-mm mortar whose gunners had trained to provide close-in fires. At Bella were two 120-mm mortars and at Camp Blessing were a pair of 155-mm howitzers, each capable of ranging the area around the Ranch House. There was also a pair of Air Force A-10 close support jet aircraft and Army Apache attack helicopters on call for missions in the area. While the mortars and howitzers were immediately available, the fixed and rotary wing assets were stationed at Bagram Airbase and Jalalabad and required between 30 minutes and an hour to be on station.

In the TF *Rock* area of operations, 21 August 2007 was a relatively quiet day but this calm was deceptive. The enemy was well prepared to attack the outpost. Aranas insurgent leader Hazrat Omar would personally lead the attack which would include both the employment of supporting fire positions all around the outpost and an assault force which would mass against the southeastern corner of the perimeter where the Afghan troops were positioned. Enemy intelligence on the outpost and its garrison was extensive. After the battle a detailed schematic of the outpost's setup was discovered on a captured camera.

On the morning of 22 August 2007, there were 20 members of the 1st Platoon, C Company, 2-503 IN, and the supporting mortar squad at the Ranch House position. This force included one officer, eight noncommissioned officers and 11 enlisted men. A medic from the brigade medical company (B Company, 173d Support Battalion) and a forward observer were also attached to the platoon. Among the paratroopers were 13 who had not yet seen combat. The outpost commander was the platoon leader, 1LT Matthew Ferrara. Ferrara's senior NCO was Weapons Squad Leader SSG David Dzwik. Dzwik was assisted by SSG Erich Phillips, who led the 60-mm mortar squad. In addition to the Americans there were about 22 ANA soldiers and about 45 ASG fighters, including those at the distant OP. Advising the ANA detachment was a small US Marine embedded training team (ETT). With the rest of Ferrara's platoon providing the garrison at COP Bella, company commander Myer, rotated between Camp Blessing, where the company's 2d Platoon was located, and Bella. On 22 August, Myer was at Bella. Typically the squads rotated between patrolling the areas around the COP and manning the defenses of the outpost. Platoon headquarters and mortar personnel often augmented the patrolling units.

Before dawn that day at the Ranch House, all was quiet. The defenders were manning their posts at the routine security level of 25 percent, which meant at each position an average of one person was awake at any given time. Not expecting a dawn attack, stand to (i.e. 100 percent alert) procedures were not in place. The ASG at Post 5 were awake and conducting morning prayers per Islamic custom. Although the defenders routinely sent out patrols at irregular intervals night and day, no patrols were out during this predawn period.

At 0454 hours, with dawn approaching, the quiet was suddenly broken. At the Ranch House a force of four or five insurgents appeared on the hillside about 100 meters south and east of the outpost in the rugged terrain between the outpost and Aranas. The enemy was dressed in BDU-style clothing similar to that which had been issued to ASG personnel. The attackers focused their fires on the ASG position (Post 5), Post 3, and the TOC position. As it was in the most vulnerable position, the garrison at Post 3 had the least time to get ready once the attack began. Accordingly the enemy was able to concentrate against the position with only the return fire of the M240 machine gun from the duty soldier in the tower. In rapid succession four RPG rounds struck the post virtually destroying it and damaging the Mk-19 grenade launcher and M240 posted there. The paratrooper on duty in the tower, SPC Jeddah Deloria, survived in the wreckage wounded but still capable of fighting. With the radio destroyed at Post 3, the NCOIC there, SGT Carlos Gonzales, sent SPC Charles Bell to Post 2 to report and continue the fight.

Gonzales himself was wounded shortly thereafter and with the enemy approaching the ruins of the post, he too withdrew to Post 2. Before departing, however, he told Deloria to lie still so the enemy would not notice him and that he'd send soldiers to get him. At Post 2, Bell reported the situation at Post 3 to SPC Sean Langevin, the soldier on duty there, and began firing that post's Mk 19 grenade launcher at the insurgents now closing in on Post 3. RPG rounds started landing near Post 2 hindering the ability of SGT John Relph, the NCO in charge, and PFC Adam Spotanski from moving up to the platform from their sleeping quarters. Langevin provided covering fire and the duo managed to reach the post platform and begin firing in the direction of the attackers. After Gonzalez's arrival at the post, Relph and Spotanski attempted to move to Post 3 to rescue Deloria but were unable to do so because of the volume of enemy fire.

After the brief opening volley, the ASG and ANA elements located at and near Post 5 broke contact and withdrew to the center of the perimeter. Many of the ASG men withdrew completely, retreating into Aranas and the countryside away from the enemy positions. The withdrawal left the southeastern portion of the perimeter undefended. The attackers did not

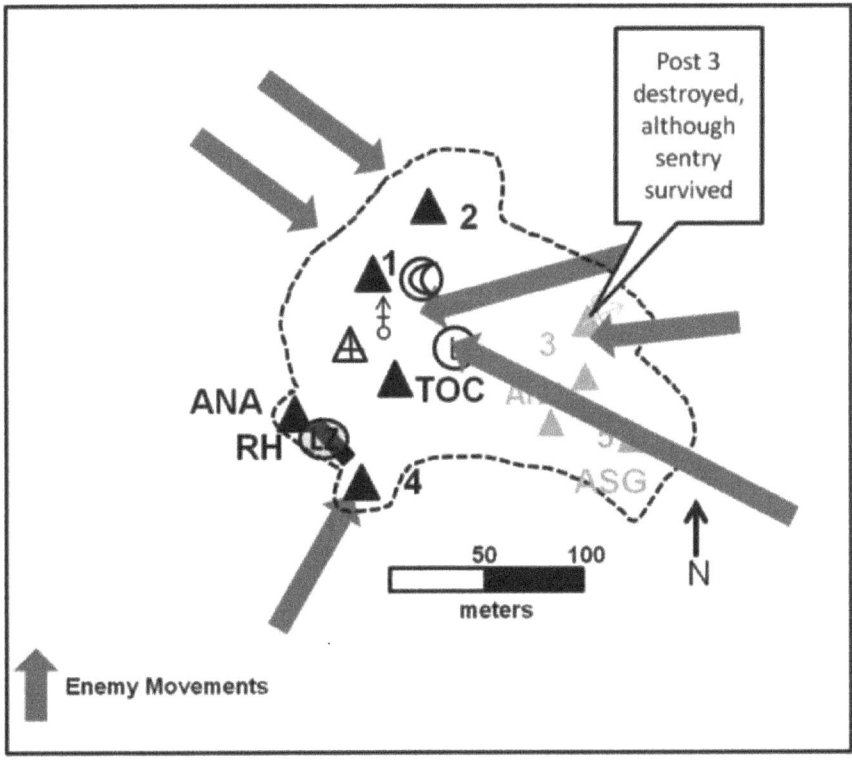

Figure 3. The Initial Attack on the Ranch House.

immediately take advantage, initially preferring to mass fires against all the positions in the outpost before advancing across the now abandoned portions of the perimeter. The American defenders fought back vigorously, firing off all prepositioned Claymore mines, throwing hand grenades and returning the enemy fire with machine guns and volleys of 40-mm grenade rounds.

With all positions under heavy fire, the American defenders spent the early part of the fight returning small arms fire and setting off Claymore mines. Only Post 1 remained relatively unscathed although its garrison also returned fire in all directions. In the TOC, 1LT Ferrara immediately contacted Captain Myer at Bella via FM and satellite radio. Myer promptly requested close air support from the battalion TOC at Camp Blessing. Realizing there was at least a 30-minute period before such support could arrive on the battlefield, Ferrara also requested 120-mm mortar fires from Bella against enemy forces aiming RPGs and machine gun fires at the TOC. The attack had started so suddenly that the enemy was already too close to fire the 155-mm field artillery guns at Camp Blessing due to the danger of fratricide. The 120-mm mortars at Bella posed similar risks and Ferrara and Myer aimed their fires at ridgelines 200-300 meters away from the Ranch House against more distant enemy positions and presumed rear echelons of the advancing insurgent forces. The battalion commander, LTC William Ostlund, alerted the unit quick reaction force (QRF), two squads from A Company, stationed in the Pech Valley about 16 kilometers (ten miles) south of Aranas. The squads would be airlifted, with one earmarked to reinforce Bella which both Myer and Ostlund feared would be attacked, and the other to the Ranch House position.

As the intense initial firefight continued, a force of about 20 insurgents took advantage of the destruction of Post 3 and the abandonment of Post 5 to advance through the newly created gap inside the perimeter of the outpost in an effort to overwhelm and overrun the defenders. In this they were partially assisted by their wearing of BDU-style uniforms that prevented the Americans from distinguishing the assaulters from ASG troops. As the enemy forces closed in on the northern positions, their fires became more intense aided by a stockpile of RPGs captured at Post 5. These fires soon damaged the antenna array located on the roof of the TOC bunker, cutting a wire to the satellite dish and otherwise damaging the remaining antennas. Ferrara and his radio operators, SGT Conrad Begaye and SPC Kain Schilling, lost communications with the outlying posts and with CPT Myer at Bella. After about four minutes, Ferrara was able to reestablish contact with Myer by moving his FM radio outside the TOC bunker and using a smaller undamaged antenna. Begaye and Schilling covered their platoon leader as he continued to relay updates and requests for support to their company commander.

SSG Erich Phillips was an experienced noncommissioned officer with a background as a scout and as a mortar gunner. At the Ranch House, he technically served as the leader of the small 60-mm mortar squad but Ferrara and Dzwik depended upon him for his knowledge and professionalism, frequently using him as a patrol leader. At the start of the fight, Phillips was asleep in his quarters near the mortar position. He quickly got into action and marshaled an *ad hoc* group from around the mortar and TOC positions to defend the new line to the south. SPC Jason Baldwin, a mortarman, and SPC Kyle White, the platoon radio operator, had been reinforcing Post 1 when the battle started. With that post under the least pressure, the two men ran to the mortar area where they linked up with Phillips who had already assembled mortar gunner SPC Hector Chavez, platoon forward observer Schilling, and platoon medic SPC Kyle Dirkintis into a small reserve force he intended to use to counterattack or otherwise restore the defensive perimeter. As this force was formed, ANA and ASG Afghans fled past them from the direction of Post 5 towards the ANA post near the LZ.

Ferrara told Phillips that contact had been lost with Posts 3 and 4. The men at the TOC/mortar area could hear the sounds of firing still coming from Post 4 but Post 3 was a smoldering ruin. Word reached Phillips from Post 2 to where the garrison of Post 3 had evacuated, that Deloria was still at Post 3, probably wounded. Ferrara reported the wounding of Deloria to Myer, who immediately requested the dispatch of a medical evacuation helicopter. The enemy pressure on Phillips' group prevented an immediate rescue effort. The fire was so intense and the insurgents so close at less than 15 meters away that when Phillips and Baldwin attempted to load the 60-mm mortar, enemy machine gun and AK-47 rounds impacting around the tube made the attempt too dangerous. Instead the group began returning fire with their personal small arms.

To the southwest, Post 4 was now isolated from the rest of the defenders and was receiving fire from three sides. SPC Jeffrey Shaw had been on duty when the attack began and he initially fired to the east against the enemy elements concentrating against posts 5 and 3. Within minutes the rest of the garrison, SGT Michael Johnson, PFC Gregory Rauwolf, and SPC Robert Remmel, arrived from their sleeping area as the position began to receive fire from insurgents located to the west. While Johnson and Remmel stayed at the post, Shaw and Rauwolf moved to secondary positions to facilitate firing at the enemy approaching from the east. Enemy fire from the direction of the evacuated ASG position struck Remmel in the back. Meanwhile Shaw and Rauwolf shot two dozen 40 millimeter rounds from his M203 and Rauwolf fired 200 rounds of 7.62 millimeter M240 machine gun ammunition. The enemy to the east was soon suppressed and unable to advance from that direction towards Post 4.

The enemy pressure remained intense, however. While Rauwolf treated the wounded Remmel, Johnson took over the M240 machine gun in the tower, occasionally mixing in M203 rounds from his personal weapon. Shaw added his fire from the secondary position on the ground nearby. The volume of fire received at the tower became so large that Johnson was forced to move to his secondary position. From there he and Shaw continued to fight and were joined by five Afghan soldiers and their American advisor. During this action, Shaw received wounds in his arms preventing him from using his rifle. Johnson and the Afghan group continued to man their weapons until the end of the fight.

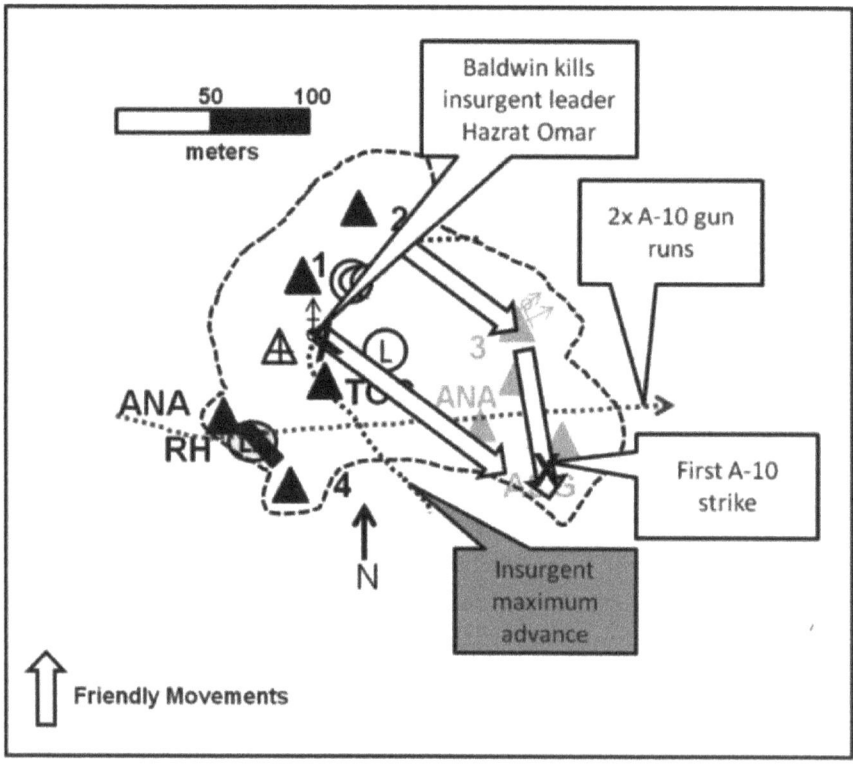

Figure 4. The Counterattack at Ranch House.

Back at the TOC, insurgent pressure continued to increase. Baldwin and Phillips threw several grenades toward the enemy at the outpost's latrine, which now marked the frontline. Mortarman Chavez informed Phillips that Gonzalez was wounded at Post 2. Phillips decided to grab the platoon medic and go to that post to check on Gonzalez. Phillips and medic Dirkintis then ran the 20-meter distance to Post 2 under enemy fire while Baldwin covered their movement by throwing volleys of grenades towards the enemy near the latrine. Once there, Dirkintis was quickly wounded in the shoulder by RPG shrapnel as insurgent fighters began closing in on Post 2. To blunt this advance, Relph and Phillips threw several hand grenades while Bell fired the post's M240 machine gun. However, Bell was soon targeted and wounded by small arms fire.

Despite his wound and after some quick first aid, Bell resumed firing toward Post 3. Langevin supported Bell by firing a squad automatic weapon (SAW) while Relph fired his M4 carbine. Langevin was slightly wounded in the leg but continued to man the SAW. Meanwhile, Relph was seriously wounded and joined Gonzalez in the bunker below the post where the two NCOs passed grenades up to Langevin who threw them at the enemy who were now quickly approaching the position.

Near Post 2, Phillips felt that Dirkintis' wounds required immediate treatment at the aid station and he began dragging the medic back towards that location while Langevin covered him from Post 2. On the way back, enemy fire became so intense that Phillips left Dirkintis in a culvert and returned to organize the defense near the TOC. Once there he sent Chavez, who had received specialized medical training, back to conduct first aid on the wounded medic. After treating Dirkintis' wounds and ensuring he was in a relatively safe, covered position, Chavez returned to the TOC area and joined the defense, providing M203 fire. Phillips had immediately deployed the small force near the TOC with a loose line between Post 2 and the TOC area to ensure that there were no gaps through which the enemy could advance.

While Phillips was gone, Baldwin had put the 60-mm mortar into action, firing it in hand-held mode, with several rounds going through the roof of the ASG post. Baldwin alternated mortar rounds with hand grenades as the insurgents continued to close on his position. Upon his return, Phillips assisted Baldwin with the mortar. The duo fired about 20 rounds at the enemy position in the perimeter breach. The nearest rounds landed only 63 meters from the Americans. The insurgents tried to rush the mortar position approaching to within ten meters before Baldwin's fire killed the leader of the local insurgent cell and the attacking enemy force commander, Hazrat Omar. The combined effect of the defenders' fire and the sudden loss of leadership stopped the insurgent advance.

The defending paratroopers expected a renewed enemy advance and hoped that close air support arrived soon. The pair of US Air Force A-10 aircraft that Myer had summoned at the start of the action arrived in the general area of Aranas one hour and four minutes after the start of the attack. With the airplanes on site, Myer passed direct communications with them to Ferrara. The A-10 is an aircraft designed specifically to provide close air support to ground forces. As such, it has armored plating to allow it to survive ground fire, a nose cannon capable of firing 65 30-mm rounds a second and up to six Maverick air-to-surface missiles mounted on its wings. With the opposing forces so close together, both Ferrara and the pilots had to be careful to avoid fratricide. The fact that the ASG post was now on fire, the result of Baldwin's mortar gunnery, aided the pilots in identifying the enemy positions. In order to verify this, the A-10s dropped several flares and a white phosphorus round on the suspected enemy positions. After receiving Ferrara's acknowledgement that the aiming points were, in fact, occupied by the insurgents, the airmen prepared to conduct gun runs.

The first A-10, codenamed HAWG-17, orbited the battle area and moved into its gun run from west to east, flying across the southern edge of the outpost firing a spray of 30-mm rounds from its nose gun from near the TOC to the ASG post. The gun run followed the same basic orientation that Baldwin and Phillips had used when firing the mortar. Ferrara observed the rounds as landing almost exactly where he wanted them to, with the nearest rounds falling within 50 meters of the TOC. Phillips saw the A-10 fly right over the TOC with the closest rounds impacting near Chavez and Dirkintis. The A-10 also placed two missiles into the ASG tower. The aerial support had the desired effect. With the enemy advance already halted by Omar's death, the intensity of the insurgents' fire immediately decreased by half.

A second gun run followed the first. This run, while along the same trajectory as the first, started closer to the TOC, at a slightly steeper angle. Shrapnel from the closest impacts slightly wounded Begaye, who was standing near Ferrara. While enemy fires had decreased, they had not stopped totally and were still intense in several places. At Post 2 Langevin noticed insurgent rounds landing within 20 meters of his position and a force of about 20 men throwing grenades at the aid station from higher ground above it.

Throughout the action, 1LT Ferrara had been calling in casualty reports, revised with updated information. Two medical evacuation (MEDEVAC) helicopters, escorted by an Apache gunship, had flown from Jalalabad to Asadabad, where it waited for the fighting to die down. Once the QRF was ready to go, it was delayed, with the MEDEVAC flight getting the priority. At Camp Blessing, the battalion supply officer (S4) prepared an ammunition resupply "speedball" bundle for aerial delivery. However, since the Ranch House defenders had actually fallen back onto their ammunition resupply point stockpiled with more than the unit's basic load, the speedball proved to be unnecessary.

After the second A-10 gun run, insurgent fire and pressure gradually died down. Ferrara now saw the opportunity to rescue Deloria buried in the ruins of Post 3. He dispatched Phillips on this mission, who took Baldwin and one of the ANA advisors with him. Instead of going directly to Post 3, Phillips stopped at Post 2 on the way to check on the wounded men there. Of the four - Gonzales, Bell, Relph and Langevin - only Gonzalez required immediate attention and Spotanski, an unwounded member of the garrison, took him to the aid station. Spotanski soon returned and when Phillips proceeded to Post 3, he took him along. Baldwin, the slightly wounded Langevin and the remaining garrison of Post 2 covered the advance. The duo reached the wrecked position with only a few enemy potshots aimed at them and found Deloria under the debris. While they dug him out, Baldwin and Langevin also moved up to the destroyed position. Deloria, escorted by Phillips, was able to walk on his own to the aid station while Baldwin, Langevin and Spotanski manned Post 3 with a squad automatic weapon as their primary armament, clearing out the wreckage at the same time. In the process of clearing out the debris, Spotanski fell and impaled himself on a pole, becoming the last American casualty of the action.

The first MEDEVAC helicopter arrived after the insurgent fire had been mostly silent for a half hour. While the enemy outside the outpost was observed evacuating their casualties, the second MEDEVAC aircraft was fired on at a distance south of the Ranch House. Meanwhile the troops moved all the wounded down to the landing zone for evacuation and the ANA soldiers cleared the portion of the outpost formerly overrun by the insurgents and reoccupied their original positions. Within minutes of the departure of the first MEDEVAC flight, the members of the QRF, a squad from A Company, and the 1st Platoon's platoon sergeant, SFC William Stockard, arrived. As planned, another squad reinforced Bella.

With the wounded and injured evacuated, the situation at the Ranch House outpost returned to its pre-battle quiet. Although the ASG had returned, the former ASG post was now manned by Americans. The attack refocused the attention of TF *Rock*'s leadership on the security of the COP. LTC Ostlund had determined before the attack that Ranch House's proximity to a center of insurgent activity did not make up for its remoteness and had planned to close the COP, moving its garrison to a new site located next to the Waygal District Center at the village of Wanat, only six miles (10 km) from Camp Blessing. The attack accelerated these plans. Because of difficulties in coordinating the establishment of a new COP at Wanat, TF *Rock* now planned on moving the Ranch House garrison to Bella until the Wanat post could be created, in which case Wanat would replace both Bella and Ranch House. Myer planned to expand Bella into a platoon-sized outpost. With the additional troops, the garrison could expand its patrolling and its interaction with the local population. COP Ranch House was evacuated on 2 October 2007. The members of the 1st Platoon moved to Camp Blessing for a brief rest and then joined the rest of the platoon at Bella.

In a later interview Phillips felt that the fight would have been a lot less desperate if the Afghans had not abandoned the ASG post:

> if the ASG and the ANA were holding their ground, we already had a QRF plan established in the event we took major contact on the FOB. I would have grabbed me two or three dudes. They weren't in that big of a firefight. We could have pushed up there and reestablished and helped them out. That wasn't the case because they ran within the first five minutes of the fight. They brought the enemy right to my front door. Within 15 to 20 meters, I'm trying to fight off 60 dudes.

The Americans at the Ranch House outpost were conducting an ongoing counterinsurgency operation that required the unit to operate in the midst of an area that had historically been a hotbed of insurgent activity. As such the half platoon had to maintain a defensible base camp that provided it an ideal location from which to interface with the local population while conducting various non-combat activities in overall support of counterinsurgency objectives. Even when the unit's QRF plan dissolved, Phillips and the other NCOs at the outpost displayed initiative and verve in their reactions to the attack, fully understanding this mission, how the defense was expected to work, and what was necessary to prevent disaster. With the NCO leadership directing the actions of the squads and sections, 1LT Ferrara was free to maintain contact with his superiors and direct fire support activities.

The Army recognized the valor of the Ranch House garrison. Phillips was awarded the Distinguished Military Cross. Ferrara and Baldwin received the Silver Star. Seven soldiers were awarded the Bronze Star Medal with V Device for their actions, with an additional five receiving the Army Commendation Medal (ARCOM) for valor. Thirteen members of the garrison were awarded the Combat Infantryman Badge, the Combat Medic Badge or Combat Action Badge, indicating the attack was their initiation into combat. During the action 11 paratroopers were wounded or injured out of a garrison of 22, a 50 percent casualty rate that dramatically demonstrates the intensity of the combat at the Ranch House.

1LT Ferrara Briefs MG Rodriguez on Ranch House Fight.
US Army photo.

After the attack, stand-to became standard procedure in C Company. This came in particularly handy a year later in the 13 July 2007 insurgent attack on a new COP built at Wanat, about ten miles southwest of Aranas. There the members of C Company's 2d Platoon were alert and on stand-to at dawn when the enemy attacked. Phillips, Dzwik and Chavez repeated their heroics in this later fight. However, several of the Ranch House defenders, including 1LT Ferrara, were no longer alive by the time of the Wanat attack. On 9 November 2007, Ferrara, Langevin, and four other Soldiers were killed in an ambush while returning to Bella from a shura. During the ambush, RTO and Ranch House veteran Kyle White distinguished himself and was nominated for the Medal of Honor. SGT Begaye, who worked radios during the Ranch House fight, was awarded a Silver Star for his actions in the fight.

The August 2007 Ranch House fight was similar yet another major attack on an outpost in October 2009. COP Keating located near Kamdesh, about 20 miles northeast of Aranas, occurred on 3 October 2009. In both cases, the outposts were projected to be abandoned soon. And similarly, the Afghan-manned section of the perimeter collapsed, resulting in close-in fighting. Both attacking and defending forces at Keating were larger than at the Ranch House. While the fighting was equally fierce, eight Americans were killed at Keating, one indication of how the war in northeastern Afghanistan had intensified since 2007.

For Further Reading

James Christ. *The Dirty First at Aranas*. Chandler, AZ: Battlefield Publishing, 2011.

Combat Studies Institute staff. *Wanat: Combat Action in Afghanistan, 2008*. Fort Leavenworth: Combat Studies Institute Press, 2010.

Sebastian Junger. *War*. New York: Twelve, 2010.

David Kilcullen. *The Accidental Guerilla: Fighting Small Wars in the Midst of a Big One*. New York: Oxford University Press, 2009.

Jake Tapper. *The Outpost: An Untold Story of American Valor*. New York: Little, Brown and Company, 2012.

The Six Principles of Mission Command

1. Build Cohesive Teams through Mutual Trust

2. Create Shared Understanding

3. Provide a Clear Commander's Intent

4. Exercise Disciplined Initiative

5. Use Mission Orders

6. Accept Prudent Risk

Mission Command in the Ranch House case

1. Build Cohesive Teams through Mutual Trust. The company and platoon were cohesive. Although it had only arrived in Afghanistan several months earlier, the members of the units had trained together in preparation for the deployment. The platoon leader and squad leaders had led their units since the start of the rotation. The unit was small enough that all the participants knew each other well. The soldiers trusted their NCOs, particularly SSG Phillips

2. Create Shared Understanding. The unit at the Ranch House outpost was conducting an ongoing counterinsurgency operation that required the unit to operate in the midst of an area that had historically been a hotbed of insurgent activity. As such the half platoon had to maintain a defensible base camp that provided it an ideal location from which to interface with the local population while conducting various non-combat activities in overall support of counterinsurgency objectives. All the members of the garrison understood the mission, the defensive set up, and the plan for a QRF if the COP was attacked. Further, they were able to respond to relatively general instructions when the Afghan-manned section of the perimeter collapsed. The unit had previously rehearsed quick reaction drills and fighting from secondary and supplemental positions. These preparations became important during the action.

3. Provide a Clear Commander's Intent. The unit at Ranch House was executing an ongoing mission as part of a counterinsurgency campaign. As such the commander's intent was to ensure that the outpost could sustain itself in order to conduct counterinsurgency operations in the Aranas area. All members of the garrison had a clear understanding of this and responded, once under attack, accordingly.

4. Exercise Disciplined Initiative. Initiative was most apparent at the platoon and squad leader levels. The company commander allowed the platoon leader to direct indirect fires and close air support based on

his understanding of the situation while he focused on ensuring those resources were available and that reinforcements were being marshaled. The NCOs at each post and, in particular SSG Phillips, did not have to be given detailed instructions in response to the enemy attack.

5. Use Mission Orders. During the action the platoon leader did not issue detailed instructions to his subordinate leaders, instead giving them general instructions while he focused on provided necessary external support. By the same token, the battalion and company level commanders did not give detailed instructions to the defenders during the action, instead focusing their efforts on ensuring fire support assets were available and expedited and that reinforcements were promptly dispatched.

6. Accept Prudent Risk. This principle is the one most apparent in the Ranch House battle. The battalion and company commander had to accept a certain amount of risk based on the extended area their units had to cover and the need to place troops among the local population in rugged terrain. However this risk was prudent because fire support assets were able to range the outpost and a quick reaction force system had been put in place to provide responsive reinforcements to the outpost. During the action this was demonstrated by the almost instant availability of distant mortar and artillery fires. While these were hampered by the proximity of the enemy, this gap was filled by the 60-mm mortar at the outpost and the mutually supporting small arms fires of the defenders. While there was a time lag in the arrival of close air support, predicated on the limited number of air assets in the theater, the A-10s arrived at the exact right time to break the back of the enemy assault, reinforcing the level of acceptable risk taken by the chain of command.

Operation NASHVILLE

Breaking the Taliban's Stranglehold in Kandahar, 2010

Anthony E. Carlson, Ph.D.

In July 2010, GEN David Petraeus, the commander of the International Security Assistance Force (ISAF) in Afghanistan, designated the 2d Brigade Combat Team (2d BCT), 101st Airborne Division (Air Assault) as the ISAF main effort. Commanded by COL Arthur Kandarian, the brigade was tasked to conduct offensive operations in Kandahar Province's Zhari District, the birthplace of the Taliban. Since 2006, when a Canadian-led task force defeated a large concentration of Taliban fighters preparing to attack the nearby city of Kandahar, the Taliban had reasserted control over Zhari. The insurgent group had assassinated key tribal elders, established a shadow government including a de facto "supreme court," and tortured political prisoners. The Taliban also had a stranglehold over commerce. On Highway 1, Zhari's major thoroughfare connecting Helmand Province to the west with the city of Kandahar to the east, the Taliban set up illegal checkpoints to collect exorbitant tolls. Drivers who refused to pay were swiftly assaulted. By 2010, with a growing insurgency on its western doorsteps, the second largest city in Afghanistan suffered from political instability and economic stagnation.

To break the insurgency's iron grip on Highway 1, COL Kandarian planned a series of coordinated operations south of the highway. The plan, christened Operation DRAGON STRIKE, involved his brigade's two maneuver battalions (1st Battalion, 502d Infantry Regiment [1-502 IN] and 2d Battalion, 502d Infantry Regiment [2-502 IN]) and its Reconnaissance, Surveillance and Target Acquisition (RSTA) squadron (1st Squadron, 75th Cavalry Regiment [1-75 CAV]) clearing the insurgent sanctuary south of Highway 1 in Zhari. In July, the ISAF's Regional Command-South (RC-South) bolstered Kandarian's combat power by assigning LTC Bryan Denny's 3d Squadron, 2d Stryker Cavalry Regiment (3-2 SCR) to the brigade.

Kandarian expected his subordinate commanders to take bold action, exercise initiative, and accept reasonable risks. His intent was to "defeat the insurgency in Zhari ... in order to secure the people, ensure Afghan FOM [freedom of movement] on Highway 1, and improve governance and development." During his brigade's rigorous pre-deployment training regimen, Kandarian emphasized decentralized leadership based on initiative. According to 2-502 IN commander LTC Peter Benchoff, Kandarian's command philosophy was akin to drawing a circle on a map, announcing task and purpose, and letting his subordinate commanders achieve the intent without slavishly following a dictated plan. The brigade commander trusted his subordinate commanders to design schemes of maneuver tailored to their tactical environments and unforeseen contingencies within the limits of his intent.

Benchoff's 2-502 IN served as the main effort for DRAGON STRIKE. Encompassing the western third of Zhari District, the battalion's area of operations (AO) included "the most volatile and kinetic area in southern Afghanistan." The AO's naturally defensible terrain favored insurgents. Eight-foot tall earthen grape rows, marijuana and poppy fields, tree-lined irrigation canals, pomegranate orchards, and a plethora of two-story mud huts scattered in the fields facilitated insurgent cover and concealment south of Highway 1. MAJ Curt Rowland, the 2-502 IN operations officer (S3), likened the irrigation canals to "World War I, trench style type defenses." Running parallel to Highway 1, the canals enabled the insurgents to move laterally on an east-west axis, using covered positions to fire 82 millimeter recoilless rifles at highway traffic. In addition, the Taliban prepared extensive improvised explosive device (IED) belts on every north-south route connecting to Highway 1, making those routes – and indeed the entire Zhari district - a tangled maze of minefields and ambush sites.

In mid-September, Benchoff planned his battalion's first offensive to clear Objective NASHVILLE, a kilometer-wide strip south of Highway 1 near Forward Operating Base (FOB) Howz-e-Madad. By establishing a foothold south of the highway, the battalion would restore commerce and occupy the violence-plagued villages of Baluchan and Pulchakhan, meeting two key tasks in COL Kandarian's intent. Benchoff's 22 September mission statement called for clearing "the vicinity of Objective NASHVILLE beginning on 25 Sep 10 in order to hold, creating freedom of movement along Highway 1 ... and safeguarding the people immediately south of the Highway." Benchoff instructed his subordinate commanders at all costs to avoid inflicting civilian casualties (CIVCAS) which would alienate the villagers living south of the highway.

Benchoff selected CPT David Yu's Bravo Company as the main effort of what was now known as Operation NASHVILLE (see Map). Yu's company would air assault into the village of Baluchan at night, search compounds of interest, meet village elders, and collect the biometric data of Afghan military age males (MAM). Two kilometers to the west, two platoons from CPT David Forsha's Alpha Company (1LT Thomas Meyer's 1st Platoon and 1LT Barrett Rife's 2d Platoon) would attack to the east of Pulchakhan, searching compounds and establishing temporary strong points (SPs). To the east of Bravo Company, two platoons from CPT Timothy Price's Delta Company (1LT Kyle Snook's 1st Platoon and 1LT Sayre Payne's 2d Platoon) and a company of Royal engineers from the United Kingdom's 1st Armored Engineering Squadron would build a new road (Route TENNESSEE) from SP Spin Pir on Highway 1 southwest to an unoccupied Afghan compound a kilometer southeast of FOB Howz-e-Madad. Benchoff intended to insert CPT William Faucher's scout platoon into that compound, designated as Outpost (OP) Dusty, by helicopter. The battalion commander believed that constructing TENNESSEE would divert insurgents away from Baluchan, isolate them to the north, and allow Price's element to bypass the impassable north-south routes.

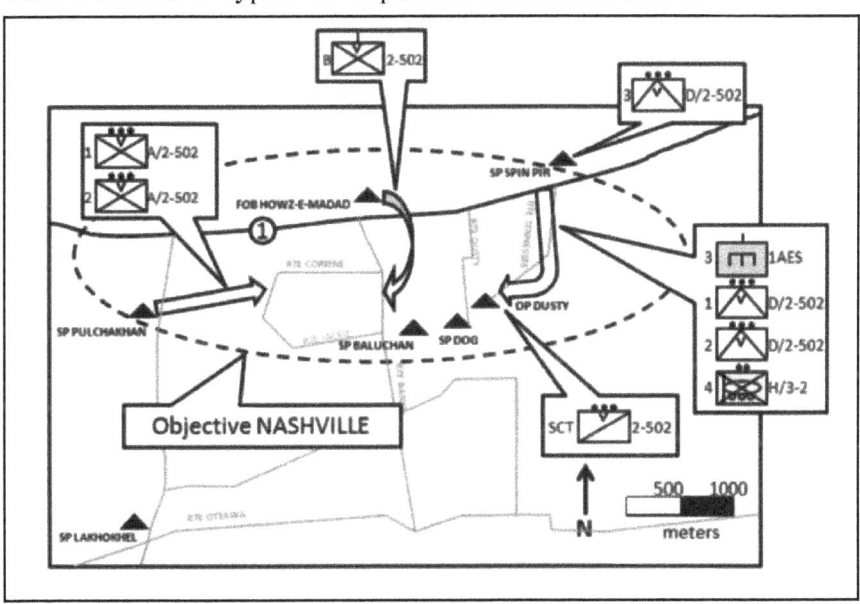

Map. Scheme of Maneuver – Operation NASHVILLE.

On the verge of the operation, Benchoff was confident that his company commanders and noncommissioned officers (NCOs) shared an understanding of the operation's purpose, potential problems, and the environment south of Highway 1. The extensive combat experience of his subordinate commanders *and* their collective understanding of his and CSM Troy Henderson's tactical standard operating procedure (TACSOP) constituted the basis of his confidence. Benchoff described the battalion TACSOP, which was understood all the way down to the team leader level, as his "commander's intent for the close infantry fight." Zhari's restrictive terrain put a premium on flexible, aggressive small unit leadership and decentralized command and control. "In difficult terrain where you have isolated units," said Benchoff, "you've got to have that leadership with the drive and motivation and understanding of the intent and the desire to seek out opportunities to make success."

NASHVILLE commenced on the night of 26 September. CPT Yu's Bravo Company was inserted by air just to the northwest of Baluchan. During the next two days, the company searched compounds suspected of IED production, held *shuras* [meetings] with village elders, and entered the biometric data of Afghan MAM into a digital database. Remarkably, insurgent resistance was negligible. CPT Forsha's two Alpha Company platoons also faced little enemy contact as they advanced to the east.

As Alpha and Bravo Companies moved towards their objectives, CPT Faucher's scout platoon was inserted at OP Dusty before sunrise. The scout platoon consisted of three reconnaissance teams of five or six soldiers and a sniper section divided into three teams of three Soldiers (a spotter, a sniper, and a security man). A seven-man Afghan National Army (ANA) reconnaissance element accompanied the scouts. Faucher's soldiers used C4 demolition charges to clear the compound of IEDs and set up a defensive perimeter. Encircled by a four-meter high mud wall, the elevated compound offered a panoramic view of the surrounding terrain and nearby roads.

At OP Dusty, a kilometer to the east of Baluchan, the scout platoon encountered significant enemy contact. From positions concealed in Zhari's maze of tree-lined irrigation canals, grape rows, and abandoned mud compounds, the insurgents initiated eight daytime firefights with machine gun bursts and rocket propelled grenade (RPGs) volleys. Faucher countered by radioing for multiple Apache attack helicopter gun runs, two dozen 120 mm mortar fire missions, and 13 155 mm artillery fire missions. Air Force F-16s also employed three Guided Bomb Unit (GBU), 38 Joint Direct Attack Munitions (JDAMs), and five GBU 12s, all within 300 meters of the OP. On the next day, Faucher's scouts faced almost continual enemy pressure in the form of small arms and rocket fire. While the attack helicopter support and indirect fire support made the risk Benchoff accepted in sending the scout platoon into OP DUSTY reasonable, the insurgents maintained their intensity.

As the scouts faced the threat at OP Dusty, CPT Price's team departed SP Spin Pir just north of Highway 1 before sunrise. In the vanguard were the Royal engineers, followed in order by 2d and 1st Platoons. The Royal engineers used two 62,500-kilogram Trojan Armored Vehicles (AVRE) to construct Route TENNESSEE through the dense terrain. As the AVREs smashed through the foliage, the enemy opened fire on Price's soldiers with machine guns and RPGs. The fire, as well as the dense terrain, slowed the movement toward OP DUSTY. To keep up the advance, Price's team called for several attack helicopter gun runs. "The way I saw my role was to keep [the insurgents] pinned down," explained 1LT Payne. The platoon leader worked feverishly to "pinpoint exactly those muzzle flash[es] … [and relay them] to the aviation assets, to the helicopters and they [were] my maneuver element because they can sweep across the objective." As the sun went down, Price's team halted and established a defensive position at the first irrigation canal 500 meters south of SP Spin Pir. The Americans had sustained just a single casualty, 1LT Snook had triggered a pressure plate IED that ripped off one of his feet. He was medically evacuated.

Price's Soldiers in contact along Route TENNESSEE.

Photo Courtesy of SGT Brandon Haggerton.

The enemy's stubborn resistance along Route TENNESSEE and at OP Dusty surprised LTC Benchoff. Initially, he anticipated that Baluchan would see the heaviest fighting but he now realized that Price needed reinforcing. Benchoff therefore attached a section of M1128 Mobile Gun

System (MGS) Strykers from 4th Platoon, Hawk Company, 3-2 SCR (4/H/3-2 SCR) to support Price. The 4/H/3-2 SCR section had been held in reserve at FOB Howz-e-Madad. Armed with 105-millimeter cannons, each MGS carried 18 rounds and could apply overwhelming precision fires to support infantry. Knowing that Price had served as a Stryker platoon leader during a previous combat tour in Iraq, Benchoff gave him freedom of action in employing the two MGSs.

When the MGS section arrived, Price briefed his plan. He explained that the movement of the Royal engineers and his two infantry platoons slowed to a crawl as insurgents massed fires on the exposed column. Price directed the MGS commanders to fire canister round volleys into the wood lines where insurgents were perched. Packed with nearly 1,000 ball bearings that fanned out in a shotgun-like pattern, the canister rounds were deadly effective against personnel targets. The presence of the MGS section had an immediate effect on the pace of the column's advance.

SGT Brandon Haggerton's M1128 MGS fires in support of Price's Soldiers.

Photo Courtesy of SGT Brandon Haggerton.

As the movement's tempo increased, Price faced a critical command decision. According to the Delta Company commander, the situation of CPT Faucher's scout platoon at OP DUSTY had "escalated and they basically became pinned down." The enemy was inching closer and closer to the scouts with each successive assault. Unless the Taliban fighters surrounding OP Dusty were defeated, Price recognized that the battalion's mission to clear Objective NASHVILLE would be seriously delayed.

Weighing his options, Price chose a bold solution. He decided to conduct a hasty attack down Route TENNESSEE toward OP DUSTY. The MGS section would move flanking either side of Price's command Mine Resistant Ambush Protected-All Terrain Vehicle (M-ATV), trapping or killing the insurgents in the 300 meters separating the OP and Price's forces. The two infantry platoons would maintain their current positions during the attack. As the assault began, Faucher relayed the location of six large insurgent positions located in compounds encircling the OP. Unfortunately, the dense vegetation negated the MGS's thermal imaging targeting, preventing the MGS commanders from pinpointing exact insurgent locations. In response, Price ordered his crew to fire the M-ATV's .50 caliber machine gun to mark the insurgent firing positions. He then ordered the MGS commanders to advance and fire on the marked locations in a dramatic show of force.

The two MGSs attacked towards the compounds, unleashing a barrage of High Explosive Anti-Tank (HEAT) and High Explosive Plastic (HEP) rounds. The MGS crews then methodically moved from compound to compound, blasting holes through doors and mud walls at point blank range. The assault ended only when the MGSs ran out of ammunition. Enemy resistance then evaporated. Inside of the compounds, Price's soldiers later discovered fresh blood splatter and trails, indicating the fate of dozens of Taliban fighters. "The arrival of the MGS on scene in the vicinity of OP Dusty completely ended the engagement and resulted in the enemy withdrawing from [Obective NASHVILLE]," Price explained. The bold decision paid off. NASHVILLE culminated with the 2d BCT establishing a foothold south of Highway 1 that it maintained and expanded throughout the remainder of the deployment.

According to LTC Benchoff, Operation NASHVILLE sharply reduced violence on Highway 1 near FOB Howz-e-Madad. NASHVILLE was just one of dozens of operations launched as part of DRAGON STRIKE but its success struck a major tactical and symbolic blow to the Taliban especially because of the area's proximity to Taliban leader Mullah Omar's home village of Sangsar a couple of kilometers to the south. Indeed, by mid-October, Taliban commanders complained to *New York Times* correspondents that the brigade's deliberate combined arms attack south of Highway 1 had "routed" their fighters and loosened the insurgency's stranglehold on Highway 1, depriving the insurgency of critical funding. Tactical victories do not always produce strategic success, but NASHVILLE showed that small combined arms teams, led by empowered leaders eager to take advantage of battlefield opportunities, could defeat determined adversaries in difficult terrain.

For Further Reading

Anthony E. Carlson, and Matt M. Matthews. *The Battle for Kandahar: The US Army in Operation DRAGON STRIKE, 2009-2010*. Ft. Leavenworth, KS: Combat Studies Institute Press, forthcoming.

Carl Forsberg. "Counterinsurgency in Kandahar: Evaluating the 2010 Hamkari Campaign." *Afghanistan Report* 6. Institute for the Study of War, December 2010.

Kevin M. Hymel. "Trapping the Taliban at OP Dusty: A Scout Platoon in Zhari District." In *Vanguard of Valor: Small Unit Actions in Afghanistan*, 157-78. Ft. Leavenworth, KS: Combat Studies Institute Press, 2012.

The Six Principles of Mission Command

1. Build Cohesive Teams through Mutual Trust

2. Create Shared Understanding

3. Provide a Clear Commander's Intent

4. Exercise Disciplined Initiative

5. Use Mission Orders

6. Accept Prudent Risk

Mission Command in the Operation NASHVILLE case

1. Build Cohesive Teams through Mutual Trust. The majority of LTC Benchoff's subordinate leaders were experienced combat veterans. For thirteen months prior to deployment, the battalion's company commanders, platoon leaders, and NCOs trained together using Benchoff's and CSM Henderson's TACSOP as a guide. COL Kandarian praised how the TACSOP fostered small unit cohesiveness and mutual trust: "[Benchoff and Henderson] had a standard operating procedure that they trained and taught which is called 'How We Fight,' and it was very focused on empowering fire team leaders, squad leaders. It wasn't just an SOP sitting on a shelf, it was an SOP that was known and understood down through team leader level."

2. Create Shared Understanding. Both the brigade and battalion commanders clearly articulated their intents, objectives, and key tasks to subordinate commanders and leaders. They conveyed that the brigade's overall objective was to reduce insurgent attacks on Highway 1 by clearing and holding a kilometer-wide swath of territory south of the highway. Moreover, Kandarian and Benchoff cultivated a collective understanding that civilian casualties were to be avoided at all costs because they jeopardized mission success, a consideration that CPT Price weighed before launching the hasty attack down Route TENNESSEE.

3. Provide a Clear Commander's Intent. Kandarian and Benchoff made subordinate commanders aware of their intent by writing general commander's intent statements that succinctly described each operation's overall purpose, tasks, and desired outcomes. The overriding purpose of Operation DRAGON STRIKE was to "ensure Afghan freedom of movement on Highway 1" in order to improve commerce and governance. Trusting the judgment of their subordinates, the commanders did not micromanage tactical fights and maximized opportunities for small unit leaders to act independently and seek out opportunities for success.

4. Exercise Disciplined Initiative. The restrictive, dense terrain south of Highway 1 led to decentralized command and control as small units were often isolated from one another. Small unit commanders were forced to exercise independent judgment during ambiguous and urgent situations and quickly adapt to changing tactical circumstances. An example was CPT Price's decision to launch a hasty "thunder run" in order to relieve the scout platoon and destroy insurgent positions near OP Dusty. Even more significantly, CPT Price exercised *disciplined* initiative by ensuring that the MGS section did not inflict a single civilian casualty. Killing civilians would have jeopardized the battalion's ability to improve local governance, security, and earn the trust of Afghans living south of Highway 1.

5. Use Mission Orders. Benchoff's NASHVILLE mission order was broad; it entrusted subordinate units with responsibility for decision making at the point of action. Intensive pre-deployment training and a low turnover in key leaders reinforced mutual trust between the battalion commander and his subordinates. "Mission command is great, but it can't be a bumper sticker," Benchoff explained. "It's got to be deeply embedded in the culture of a unit—otherwise it doesn't work."

6. Accept Prudent Risk. The decision to use air assaults and cut new roads was calculated to avoid the heavily mined north-to-south routes, to surprise insurgents, and to make the enemy "fight in two directions." These operations did include inherent risks due to insurgents' linear freedom of movement and knowledge of how to use the dense terrain to their advantage. Nevertheless, the reward—avoiding heavy IED casualties and vertically enveloping insurgents—was worth the risk. Benchoff also mitigated the risk to his Soldiers attacking down Route TENNESSEE by attaching 2-502 IN's only M1128 MGS section to CPT Price. Although the high density of IEDs and presence of Taliban dismounts made the decision risky, Benchoff believed that the tactical significance of seizing NASHVILLE's southern perimeter outweighed the MGS's deliberate exposure to potential injury. The destroying of enemy positions outside of OP Dusty and the disruption of insurgent command and control validated Benchoff's prudent risk taking.

About the Contributors

Richard V. Barbuto, Ph.D.

Professor Richard V. Barbuto is a retired armor officer. He is currently serving as the deputy director of the Department of Military History at the U.S. Army Command and General Staff College. Dr. Barbuto writes and lectures extensively on the U.S. Army in the War of 1812.

Anthony E. Carlson, Ph.D.

Anthony E. Carlson holds a Ph.D. in History from the University of Oklahoma. He currently serves as an historian on the Research and Publications Team at the Combat Studies Institute and an adjunct Assistant Professor of History for the US Army Command and General Staff College. His publications include works on Progressive Era US water and flood control policy, public works, and the antebellum Army Corps of Topographical Engineers. He is also the coauthor of a forthcoming book on the US Army's "surge" in southern Afghanistan, 2010-11.

Mark T. Gerges, Ph.D.

Dr. Mark T. Gerges is an associate professor of military history at the US Army Command and General Staff College at Fort Leavenworth. A retired armor lieutenant colonel, he has taught at the United States Military Academy at West Point before obtaining his Ph.D. from the Institute on Napoleon and the French Revolution at Florida State University. He has published papers on British cavalry doctrine, the cavalry under the Duke of Wellington in the Peninsula War, and is working on a manuscript on the command and control of the mounted arm under Wellington.

Kendall D. Gott

Kendall D. Gott retired from 20 years with the US Army as a Military Intelligence officer, and his combat experience consists of the Persian Gulf War and Operation Desert Fox. A native of Peoria, Illinois, Mr. Gott, received his BA in history from Western Illinois University, and a Masters of Military Art and Science from the US Army Command and General Staff College. Prior to returning to Leavenworth, Kansas he was an adjunct professor of history at Augusta State University and the Georgia Military College. He has authored a number of publications, has appeared on the Battle of Mine Creek episode on the History Channel, and is featured on the Fort Donelson film for the National Park Service. He also appeared on C-Span discussing the 150th anniversary of that battle. Ken joined the staff of the US Army Combat Studies Institute, at Fort Leavenworth, Kansas in October 2002, where he is now the Senior Historian.

Colonel Thomas E. Hanson, Ph.D.

Colonel Hanson served three tours in Iraq and commanded the 2d Battalion 353d Infantry Regiment at Fort Polk, Louisiana from 2009-2011. He is a graduate of the US Army Officer Candidate School and holds a Ph.D. in history from The Ohio State University.

Gregory S. Hospodor, Ph.D.

Dr. Gregory S. Hospodor is an associate professor in the Department of Military History at the United States Army Command and General Staff College, Fort Leavenworth, Kansas. He is a graduate of the College of William & Mary, the University of Mississippi, and Louisiana State University, where he completed a dissertation on the Mexican War, 1846-1848. Since joining the CGSC faculty in 2008, Greg has served as, among other things, author of the Clausewitz and Jomini lessons in the College's military history curriculum, assistant director of the Department's staff ride program, and staff group advisor. The Department of Military History named him its Teacher of the Year for 2011.

Kevin M. Hymel

Kevin M. Hymel holds an MA in History from Villanova University. He is the author of Patton's *Photographs: War As He Saw It* and coauthor of *Patton: Legendary World* War II Commander, with Martin Blumenson. Before serving on the Afghan Study Team at the Combat Studies Institute, he worked for a number of military and military history magazines as a researcher, editor and writer.

John T. Kuehn, Ph.D.

Dr. John T. Kuehn is a Professor of Military History at the US Army Command and General Staff College at Fort Leavenworth. He retired after 23 years as a naval aviator with the rank of commander in 2004. He was awarded his Ph.D. in History in 2007 by Kansas State University. He is also a distinguished graduate of the Naval Postgraduate School with a degree in Systems Engineering. He has published numerous articles, reviews, editorials, and two books—*Agents of Innovation* (Naval Institute Press, 2008) and *Eyewitness Pacific Theater* (with Dennis Giangreco, 2008). He was recently awarded a Moncado Prize from the Society for Military History in 2011 for "The US Navy General Board and Naval Arms Limitation: 1922-1937." He is also an adjunct professor for the Naval War College Fleet Seminar Program and with the MA in Military History (MMH) and MA in History Programs at Norwich University, earning the MMH Faculty Member of the Year Award in 2010-2011.

John J. McGrath

John J. McGrath has worked for the US Army Combat Studies Institute (CSI) as an Army historian since December 2002. He also served as an Army historian for over four year at the US Army Center of Military History from 1998 to 2002. He has worked for or served in the United States Army since 1978. Mr. McGrath worked full-time for the Army Reserve in Massachusetts for 15 years, both as an active duty reservist and as a civilian military technician. While in the former capacity in 1991 he served in Saudi Arabia with the 22d Support Command during Operation DESERT STORM as the command historian. He has a BA from Boston College, an MA from the University of Massachusetts at Boston, and is a PhD history candidate at Kansas State University. Mr. McGrath is the author of numerous articles and military history publications, including *Theater Logistics in the Gulf War*, published by the Army Materiel Command in 1994, and seven books published by the CSI Press, the most recent being 2010's *Wanat: Combat Action in Afghanistan, 2008*. He also has contributed to several anthologies produced and published by CSI and has been the general editor of two collected works, the most recent being *Between the Rivers: Combat Action in Iraq, 2003-2005*, published in 2012.

Nicholas A. Murray, Ph.D.

Dr. Nicholas Murray is an associate professor in the Department of Military History at the US Army Command and Staff College. His book "The Rocky Road to the Great War" (Potomac Books) is due out in 2013, along with an edited book. "Pacification: the Lesser Known French campaigns (CSI). He recently published 'Officer Education: What Lessons Does the French Defeat in 1871 Have for the US Army Today?' in the Small Wars Journal. The author's current research focus is on the use of decision games and Tactical Exercises without Troops to teach decision making to officers.

Donald P. Wright, Ph. D.

Donald P. Wright is the chief of the Research and Publications Team at the Combat Studies Institute (CSI)and holds a Ph.D. in History from Tulane University. Along with editing a number of works on contemporary US Army operations, Wright co-authored "On Point II: Transition to the New Campaign, The US Army in Operation IRAQI FREEDOM, May 2003–January 2005" and "A Different Kind of War: The US Army in Operation ENDURING FREEDOM, October 2001 - September 2005."

www.ingramcontent.com/pod-product-compliance
Lightning Source LLC
Chambersburg PA
CBHW050459110426
42742CB00018B/3311